Steele's Guide to Scientific, Technical and Medical English

First edition

Steele's Guide to Scientific, Technical and Medical English

Dr S. Steele

AMp

Academic Medical Press

Published by Academic Medical Press, a division of Academic Medical Consulting.
Nottingham, UK.

Academic Medical Consulting.
Ebury Road, Carrington, Nottingham NG5 1BB
Somniare audemus

First published 2011

ISBN 978-0-9566443-0-5

Further copies can be obtained from: http://www.lulu.com
http://www.amazon.com

ACADEMIC
MEDICAL
CONSULTING

Preface to the first edition

Steele's Guide to Scientific, Technical and Medical English is intended for intermediate and post-intermediate EFL students. It arose out of the author's Oxford experiences of teaching English to scientists, technicians and medical professionals. There were common areas of difficulty that were not adequately addressed by the accessible English language books. Such issues remain and include:

i) Few exercises that resonate with students that have a scientific, technical or medical background.
ii) Insufficient attention to words, phrases and constructions that are actually used by these disciplines in the workplace.
iii) The lack of an accessible reference guide for such students.

It is hoped that the publication of this guide will begin to meaningfully address these issues.

The author is an experienced EFL/ESP teacher, in addition to being a trained scientist and a doctor. Understandably, he is well placed to appropriately address the needs of such students. Furthermore, in producing this guide he has been heavily and directly influenced by the requirements of this collection of students. Accordingly, the author would particularly like to thank his students and colleagues at Oxford (Oxford Intensive School of English), Nottingham and Warwick for their invaluable help in the preparation of this guide.

This guide has been organized into 47 units; each of which contains learning material that is relevant to scientists, doctors and other users of technical disciplines. Similarly, the learning points have been selected to address common areas of linguistic difficulty. The matching exercises have been selected to demonstrate constructions and vocabulary used in Physical, Biological, Chemical and Medical disciplines.

Oxford, August 2011 Dr S. Steele

CONTENTS

UNIT 1: PHRASAL VERBS (ACT – BREAK)

There are thousands of phrasal verbs in English and they are often used in natural English speech. It is an advantage to be able to recognize these verbs, to understand them and to use them. The following are verbs that students can expect to meet in the scientific/medical context. Phrasal verbs are usually neutral (**N**) or informal (**I**) in social formality. The verbs are all regular unless stated otherwise. When phrasal verbs have an object, the object can often be in one of two positions:

(A) *He brought down* **the price.** Or (B) *He brought* **the price** *down.*

(To bring down = To reduce)

However (A) is the more common construction.

Act up (N): To have irregular or imperfect function. **Act up** usually refers to machines.
"My personal computer **is acting up** - just when I need it most."
"I hope that the engineer will be able to fix the spectrometer; it's **been acting up** all week."
"This CT scanner is beginning **to act up**, perhaps it is time to replace it."
"The patient says his knee is **acting up**."

Account for (N): To explain/to rationalize.
"How do you **account for** these interesting results?"
"The hydrophobicity of the molecule could **account for** its low water solubility."
"It has not been possible **to account for** his post-operative fever."

Add up (N): (1) To find the sum/total of (2) To be logical/make sense.
"Could you **add up** these figures?"
"I was **adding up** the maintenance costs of the department."
"The experimental evidence **adds up** well to support our conclusions."
"His argument did not **add up** so I disagreed with his point of view."
"Ask the nurse to **add up** his urine output for the last 24 hours."

Be in on something (N): (Be, was/were, been + in on) To be involved in.
"Dr Frederick wants **to be in on** the proposed research."
"The director **was in on** the decision to send you to Spain."
"I **wasn't in on** the process that assigned these fifth generation cephalosporins as the standard antibiotics in this hospital."

Be up (N): (Be, was/were, been + up) To be awake/not asleep.
"I have a lot of work that has to be done before tomorrow; I will **be up** all night."
"You should not **be up** so late; get some sleep and continue writing the paper in the morning."
"If we start the heart-lung transplantation now, we will **be up** all night."

Boil something down (I): To decrease in size, to summarize.
Boil down something
"The first copy of your article is too long, you'll have to **boil** it **down**."
"It will take me all week to **boil** my thesis **down** to a two page summary."

Break down (N): (Break broke broken + down) To fail/to fail to function.
"The ultracentrifuge **has broken down** so we may have to buy another one."
"I hope that the airplane's engine does not **break down**."
"This ECG machine **has broken down**."

Break through (N): (Break, broke, broken + through) To make a great or significant discovery/advance.
"He may have **broken through** in his research."
"The wave-mechanics model of the atom **broke through** to change our understanding of the atom."
"Watson and Crick **broke through** to discover the structure of DNA."

UNIT 1: PHRASAL VERBS - (ACT – BREAK) EXERCISES

Explain	**Summarize in**	**Work incorrectly**	**Fail**
Be awake	**Be logical**	**Advance**	**Be involved in**

A) Use one of the above words or phrases to rewrite the sentences below. For example:

I could not **account for** the properties of the new compound.
I could not **explain** the properties of the new compound.

1) He **was up** all night preparing his presentation.

....

2) This is a reliable piece of equipment that has never **broken down**.

....

3) The useful information in the lecture could be **boiled down** to a few sentences.

....

4) This stopwatch is **acting up**, could I borrow yours?

....

5) The results were unexpected and difficult to **account for**.

....

6) The engine must be tested until it **breaks down**.

....

7) Did you think that what he said **added up**?

....

8) We were unable to **account for** the patient's temperature.

....

Act up	**Account for**	**Add up**	**Be in on**
Boil down	**Break down**	**Break through**	**Be up**

B) Use one of the above words or phrases to rewrite the sentences below. For example:

All of the research done in the last ten years is **summarized** in a three page review article.
All of the research done in the last ten years is **boiled down** to a three page review article.

1) This pH meter **is not working** - again!

....

2) How do you **explain** the physical properties of the polymer?

....

3) I did not realize that he **was involved in** the development of the IBM PC.

....

4) Is it possible to make a significant **advance** in this area of research?

....

5) He was so excited by the work that he **stayed up** all night so that it would be finished on time.

....

6) The meeting was a waste of time; all the useful information could be **summed up** on a page.

....

7) The patient's presenting complaint was summarized in three words; severe abdominal pain.

....

8) They **were not involved in** the randomized controlled trial, so they knew nothing about the preliminary results.

....

UNIT 2: PHRASAL VERBS (BREAK - COME)

Break up (I): (Break, broke, broken + up) To finish/to come to a finish. Break up is usually used to describe the end of a College or University term.
"The term/semester **breaks up** in November."
"What are you going to do with your summer vacation/holiday after the term has **broken up**?"

Bring out (N): (Bring, brought, brought + out) To publish/produce.
"He is **bringing out** a new science textbook."
"An interesting article was **brought out** last month."
"The company will be **bringing out** a new drug next month."

Bring something down (N): (Bring, brought, brought + down) To decrease/reduce (by a person).
Bring down something
"You had better **bring down** the temperature of the incubator."
"If we don't **bring down** the humidity in the room then electrical equipment may be damaged."

Carry on (N): To continue.
"The research council may not give us enough money to **carry on** this study."
"The experiment will be left to **carry on** overnight."
"I will **carry on** researching until I am 65."
"The ward round **carried on** until 6pm."

Catch someone out (I): (Catch, caught, caught + out) To find an error or fault/to outwit.
Catch out someone
"Dr Smith was obviously irritated to have been **caught out** in front of so many colleagues; he tried to make his mistake appear unimportant."
"Be careful during the talk; don't let the Professor **catch you out**."
"He was embarrassed after being **caught out** by an expert."

Catch up on (I): (Catch, caught, caught + up on) To review/to bring someone up to date with the
Catch up with latest information.
"I need to **catch up on** my reading about AIDS."
"She spent the evening **catching up with** the news."
"She spent the evening **catching up on** the news."

Come across (I): (Come, came, come + across) To give the impression of/to be understood.
"He **comes across** as a very competitive professional."
"The lecture on thalidomide did not **come across** well."

Come across someone/something (N): (Come, came, come + across) To meet by chance/to notice.
"I was surprised to **come across** Dr Keen at the conference."
"While doing some background reading I **came across** something that might help you to deduce the drug mechanism."

Come down (N): (Come, came, come + down) To decrease\to fall.
"The atmospheric temperature has not **come down**."
"Has the reactor temperature **come down**?"
"If you cool this material its resistance **comes down** so much that it becomes a superconductor."

UNIT 2: PHRASAL VERBS - (BREAK – COME) EXERCISES

Break up	**Bring out**	**Bring down**	**Carry on**
Catch out	**Catch up on**	**Come across**	**Come down**

A) Use one of the above words or phrases to rewrite the sentences below. For example:

The examinations are scheduled for two weeks before the term/semester **ends**.
The examinations are scheduled for two weeks before the term/semester **breaks up**.

1) The carnivore **continues** to chase the herbivore until the herbivore is captured.

....

2) He enjoys **finding fault** with bad speakers at conferences; he wants to appear clever.

....

3) The geological expedition will begin two weeks after the term/semester **finishes**.

....

4) We expect that the cost of software will **decrease** over the next five years.

....

5) I really should find time to **review** the new genetic technology.

....

6) The operator can **reduce** the temperature of the nuclear reactor.

....

7) How many publications has she **produced** this year?

....

8) While reading *Nature* I **noticed** an interesting article.

....

B) Rewrite the following sentences and keep the same meaning. For example:

The development meeting **broke up** without progress.
The development meeting **ended** without progress.

1) Mendax pharmaceuticals have **brought out** a new catalogue.

....

2) We must **bring down** the temperature of the furnace by 200°C.

....

3) I can't **carry on** this line of research without further funding.

....

4) Although the speaker was well prepared, he was **caught out** by penetrating questions.

....

5) I need to **catch up on** the latest developments in solid-state physics.

....

6) During the survey we **came across** three new species.

....

7) During the autumn the temperature and humidity **come down** a little.

....

8) **Carry on** giving the patient intravenous antibiotics for two more days.

....

9) We need to **bring down** your blood pressure to 130/90 mmHg.

....

UNIT 3: PHRASAL VERBS (COME - DIE)

Come out with (N): (Come, came, come + out with) (1) To produce information verbally.
Come out (2) To be published.

"She **comes out with** some brilliant insights."
"You never know what's going to **come out** at the seminars."
"That article **came out** two years ago."

Come under (N): (Come, came, come + under) To be categorized/to be classified/to be under.

"The safety of clinical trials **comes under** the authority of the Board of Ethics."
"The study of herbivore behaviour **comes under** the research areas of the departments of zoology and psychology."

Come up with (N): (Come, came, come + up with) To produce an idea or a suggestion.

"We need to **come up with** a conclusive experiment."
"The solution she **came up with** was superb."

Count on someone (N): To rely on someone/to depend on someone.

"We can **count on** him to produce the least experimental error."
"I think that we can **count on** Silvano to produce a good piece of work."
"My group is **counting on** me to win further funding."
"The relatives were **counting on** the surgeon being successful."

Crop up (I): To happen unexpectedly/to occur unexpectedly.

"I won't be able to come to the meeting; something **has cropped up**."
"Research never goes as you wish; something unexpected always **crops up**."

Cut back (on) something (N): (Cut, cut, cut + back on) To decrease/to restrain/hold back.
Cut something back
Cut back something

"Because of the economic situation our funding has been **cut back**."
"The hedge around the field needs **cutting back**."
"The tests indicate that you must **cut back** on your cholesterol intake."

Cut down/Cut down on (N): (Cut, cut, cut + down on) To decrease.
Cut something down
Cut down something

"We have **cut down** on particulate emissions from our cars by 20%."
"To lose weight you must exercise and **cut down** your calorie intake."
"To lose weight you must exercise and **cut down on** your calorie intake."
"If you **cut down on** your smoking you will decrease your risk of lung cancer."

Die down (N): To decrease in magnitude/intensity or size.

"Look how the radio wave signal **dies down** with increasing distance from the source."
"She waited for the applause **to die down** before continuing."
"The phosphorescence **dies down** completely within microseconds."
"The popularity of povastatin has **died down** since its side effects have become widely known."

UNIT 3: PHRASAL VERBS - (COME - DIE) EXERCISES

Come out with	**Come under**	**Come up with**	**Count on**
Crop up	**Cut back on**	**Cut down on**	**Die down**

A) Use one of the above words or phrases to rewrite the sentences below. For example:

All matters of laboratory safety **are under** the control of Dr Kent.
All matters of laboratory safety **come under** the control of Dr Kent.

1) I described the problem and she instantly **produced** five solutions.

....

2) After **producing** a brilliant idea he shouted "Eureka."

....

3) I think that we can **rely on** him to successfully complete the project.

....

4) I was fixing the VDU (screen) when another problem with the computer **occurred**.

....

5) In order for our cars to be ecologically acceptable we must **decrease** toxic emissions.

....

6) The resistance of electric cables can be **reduced** by 90% by using superconductors.

....

7) The fluorescence will **decrease in intensity** within nanoseconds.

....

8) Did anything **unexpected happen** during your neurological examination of Mrs Smith?

....

B) Rewrite the following sentences without changing the meaning. For example:

I will **carry on** repeating these experiments until the results are consistent.
I will **continue** repeating these experiments until the results are consistent.

1) During the discussion he **came out with** a new idea for a solar powered house.

....

2) That compound is likely to **come under** the heading of hydrocarbons.

....

3) Kekulé **came up with** the structure of benzene after dreaming about a snake.

....

4) Can we **count on** the Romega II sales representative to tell us the truth?

....

5) If no complications **crop up** then the rocket should arrive at the moon before 12 pm.

....

6) If we want the research budget to last for the whole year, we must **cut back** spending now.

....

7) **Cutting down** on fat consumption decreases your risk of cardiovascular disease.

....

8) The controversy about genetic technology is **dying down**.

....

9) You cannot always **count on** the pulse oximeter's measurement of oxygen saturation.

....

10) During the multidisciplinary team meeting he **came out with** a brilliant diagnosis.

....

UNIT 4: PHRASAL VERBS (DO - GET)

Do without something/someone (N): (Do, did, done + without) To manage/continue without.
"I could not **do without** my computer."
"He is such a great asset to the group that I would not want to **do without** him."
"The operating theatre cannot **do without** a good anaesthetist."

Drop out (of) (N): To stop doing something/to give up/to abandon.
"Less than 10% of the students **drop out** of this course."
"After he **dropped out** of his mathematics degree course, he took up accountancy."
"The Ferrari **dropped out** after the second circuit of the race."

Ease up/Ease off (on) (I): (1) Decrease pressure/stress/tension. (2) Slow down.
"I **eased off** the accelerator pedal of the car."
"You have been overworking, perhaps you should **ease up**."
"You have been overworking, perhaps you should **ease off**."
"It is time to **ease up on** the analgesics because we don't want you to become addicted."

Fall behind (N): (Fall, fell, fallen + behind) To make little progress/to lose to/to come second to.
"We are **falling behind** on the development schedule for this drug."
"In 2007 we **fell behind** the German group in this research field."
"Our thoracic unit has **fallen behind** the national average in its cure rate for lung cancer."

Feed something into (N): (Feed, fed, fed + in) (1) To supply to a machine.
(2) To key in. To type in.
"After he **fed** the papers **into** the shredder, they were destroyed."
"It took me an hour to **feed** all the data **into** the spreadsheet."
"I **fed** the DVD **into** the computer."

Figure (something) out (I): To deduce/reason/work out.
"Could you **figure out** how long it will take for the storm to arrive here?"
"It shouldn't take long to **figure out** a way of writing a suitable computer program."
"We need to review the patient's history and examination to **figure out** a new diagnosis."

Fit in with (N): To be compatible with, to match.
"A physicist would **fit in with** the team well."
"These results **fit in** well **with** those in Smith's article."
"The fever, malaise and weight loss **fit in with** the results of the biopsy, which confirms a vasculitis."

Follow up (N): To pursue/investigate further.
"That looks like an interesting result; maybe we should **follow** it **up**."
"After **following up** the references I found that the article was unreliable."
"Good doctors **follow up** their patients to ensure their continued good health."

Get across something (N): (Get, got, got + across) To communicate well.
Get something across
"He is good at **getting across** his point of view."
"You have to be forceful to **get** your point of view **across**."

Get along with (I): (Get, got, got + along) To work well with/to have a good relationship with.
"A good scientist must be able to **get along with** others."
"He concentrates too much on **getting along with** his colleagues and too little on his work."

UNIT 4: PHRASAL VERBS - (DO – GET) EXERCISES

Do without	**Drop out**	**Ease up/off**	**Fall behind**	**Get across**
Feed into	**Fit in with**	**Figure out**	**Follow up**	**Get along with**

A) Use one of the above words or phrases to rewrite the sentences below. For example:

He is not a good communicator and so he does not **get along with** his colleagues.
He is not a good communicator and so he does not **have good relationships with** his colleagues.

1) Is it wise for this country to **continue** without nuclear power?

....

2) We cannot afford to **abandon** the race to develop a popular electric car.

....

3) The farmers in this area should **decrease** their use of nitrates.

....

4) The Europeans are **losing** to the Americans in the race to explore outer space.

....

5) If you don't **put** the paper into the photocopier correctly, it won't work.

....

6) These results don't **agree** well with your hypothesis; you should repeat the experiment.

....

7) Can you **deduce** the solubility of this compound from its solubility product?

....

8) It is wise to **pursue** interesting observations because they may reveal something important.

....

9) The least interesting part of any clinical trial is **typing** the data into the statistics package.

....

10) The hospital cannot **manage without** an intensive care unit.

....

B) Rewrite the following sentences without changing the meaning. For example:

We cannot afford to **drop out** of the race to develop an electric car.
We cannot afford to **abandon** the race to develop an electric car.

1) If you want to finish the assignment before April, you should not **ease off** now.

....

2) If you **fall behind** on your assignments you will have to work very hard over the vacation.

....

3) **Feed in** the data by hand; there is no automatic transfer between the two computers.

....

4) A scientist with your training and experience would **fit in with** this research group.

....

5) It shouldn't take long to **figure out** a way to make this organic compound.

....

6) Her seminars are well attended because she is a good speaker who **gets** her message **across**.

....

7) He is often invited to attend conferences because he **gets along with** his peers.

....

8) Your morbidity and mortality statistics show that your performance is **falling behind** the other surgeons.

....

9) 20% of the doctors will **drop out of** this very competitive postgraduate course.

....

UNIT 5: PHRASAL VERBS (GET)

Get away with (I): (Get, got, got + away) To escape or avoid the consequences of something.
"Don't let the speaker **get away with** any careless errors."
"Are you sure that you can **get away with** that assumption in your proof?"

Get back to (N): (Get, got, got + back to) To return to an activity or person.
"I'm busy at the moment, could I **get back to** you later?"
"After lunch I **got back to** reviewing the paper but I didn't do much."
"I need to answer this bleep now and I will **get back** to the ward round later."

Get by (I): (Get, got, got + by) To manage with/to survive with.
"We will just have to **get by** with what we have."
"As the size of the grant has been reduced, we will have to **get by** with a smaller amount."
"This laboratory is not as well equipped as I would like but we **get by**."
"This newly diagnosed diabetic can **get by** without insulin."

Get someone down (N): (Get, got, got + down) To depress.
"The work was not producing statistically significant results and it was beginning to **get** me **down**."
"Nothing seems to **get** him **down**; he is always happy."
"Initially, the diagnosis of cancer **gets** most patients **down**."

Get something down (N): (Get, got, got + down) To write down.
Get down something
"The lecturer waited for the students to **get** their notes **down** and then he continued."
"His secretary gave me the details of his schedule for the day; I **got** them **down** in my personal organizer."

Get on with (I): (Get, got, got + on with) (1) To continue.
(2) To work well with/to have a good relationship with.
"All he wants to do is to **get on with** the development of a new car."
"**Getting on with** people was one of his skills."

Get out (I): (Get, got, got + out) (1) To deduce/calculate/work out successfully (often in the mathematical subjects).
(2) To publish/release.
"He'll never **get** a solution **out** to that mathematical expression."
"After **getting out** a solution to a previously unsolved problem, I felt a great sense of achievement."
"We must **get** our article **out** before they **get** theirs **out**."

Get over (I): (Get, got, got + over) To communicate effectively/well.
"You will never **get** your message **over** without visual aids."
"Good speakers seem to **get** their information **over** without effort."

Get round to (I): (Get, got, got + round to) To make time to do something/to eventually deal with.
"I'll **get round to** doing the background reading later."
"I wish that I had **got round** to faxing him before he left."
"I will eventually **get round to** writing my discharge summaries for these patients."

UNIT 5: PHRASAL VERBS - (GET) EXERCISES

Get away with	**Get back to**	**Get by**	**Get someone down**	
Get down	**Get on with**	**Get out**	**Get over**	**Get round to**

A) Use one of the above words or phrases to rewrite the sentences below. For example:

I'm in the middle of an assay, can I **return to** you later?
I'm in the middle of an assay, can I **get back to** you later?

1) A university researcher cannot **survive** with only one publication per year!

....

2) The meteorologist hoped that the bad weather would not **depress us**.

....

3) I tried to **write down** the information from the slide before the speaker replaced it.

....

4) Stop talking and **continue** with the test.

....

5) He's the sort of mathematician who can **work out** an answer to any problem.

....

6) Quantum mechanics is not an easy subject to **communicate** to the general public.

....

7) When will you **make time** to write the review article?

....

8) I can't **avoid the consequences of** showing results that are not statistically valid.

....

9) She has a post-natal depression that makes it difficult for her to **continue with** her usual way of life.

....

B) Rewrite the following sentences without changing the meaning. For example:

We will **get back** to the experiment after lunch.
We will **return** to the experiment after lunch.

1) The Research Council has decreased our funding by 20%; we must **get by** with less money.

....

2) I'm sorry I haven't **got round to** calculating the mean and standard error of these data yet.

....

3) He didn't **get** that point **over** very well in his seminar.

....

4) Let's **get on with** designing the new car - after lunch!

....

5) He applied for many computing jobs unsuccessfully but did not let rejection **get** him **down**.

....

6) When I first looked at the problem I didn't think I'd be able to **get out** a solution.

....

7) Can you **get away with** dissolving the protein in ethanol? Won't it be denatured (damaged)?

....

8) The Professor of Urology wants to discharge his clinical responsibilities as soon as possible and **get back to** research.

....

9) A paediatrician cannot **get away with** a poor bedside manner for long.

....

UNIT 6: PHRASAL VERBS (GET - GO)

Get through (N): (Get, got, got + round) (1) Complete/finish something.
(2) Communicate a message (verbally, by telephone etc).

"I was hoping to **get through** the whole exam but some sections were too difficult."
"After looking at the clock I knew that I would not **get through** all of the methodological papers before going home."
"I have tried the phone but I can't **get through**; I'll send a fax."
"You can't **get through** to him when he is so preoccupied."
"He **got through to** the Orthopaedics Department by email, and made an appointment to see a specialist."

Give something in (N): (Give, gave, given + in) To submit/to hand in.
Give in something

"**Give** this form **in** to the administration office before 5.30pm."
"If you had **given in** the application form on time, you might have had a chance of getting a job in the patent office."
"Every doctor has to **give in** their morbidity and mortality statistics today."

Give up (I): (Give, gave, given + up) To surrender/to resign/to abandon.

"I have been trying to isolate the hormone receptor for 6 months with no success; perhaps its time to **give up**?"
"She is very determined; **giving up** is not a phrase she understands."
"Perhaps it is time to **give up** on a conservative approach and try a surgical intervention."

Go ahead with (N): (Go, went, gone + ahead with) To continue with/proceed with.

"Although I was not sure that the experiment would work, I **went ahead** with it."
"The corporation has decided to **go ahead with** the development of a new passenger plane."
"The ethics committee has given us permission to **go ahead with** a clinical trial for the new heart failure medication."

Go by someone/something (N): (Go, went, gone + by) To judge by/to make a judgment by.

"**Going by** the ammeter, the current flowing through the circuit is 34 amps."
"**Going by** the expression on his face, I don't think that he was pleased."
"I have read the journal and **going by** what it says, it is likely that we are going to have to compete with another group."
"**Going by** this EEG, the patient is actually dead."

Go into something (N): (Go, went, gone + into) (1) To examine/investigate.
(2) To explain/describe.

"At first sight what he has written seems reasonable but if you **go into it** further, you will realize that his analysis is not correct."
"An investigation into the plane crash will **go into** all of the possible causes."
"I just want the risk factors for atherosclerosis, I don't want you to **go into** a description of the pathogenesis."

Go down (N): (Go, went, gone + down) (1) To have a complete breakdown of equipment.
(2) To be received.

"The university computer network has **gone down**."
"If the system **goes down** the bank will not be able to serve the customers."
"His speech did not **go down** well with the audience; they did not applaud."
"Cutting the costs of the pharmacy will **go down** very well with the accountants."

UNIT 6: PHRASAL VERBS - (GET - GO) EXERCISES

Get through	**Give in**	**Give up**	**Go by**
Go ahead with	**Go into**	**Go down**	

A) Use one of the above words or phrases to rewrite the sentences below. For example:

Chemists are **investigating** the possibility of making new organometallic compounds.
Chemists are **looking into** the possibility of making new organometallic compounds.

1) Will you be able to **finish** counting all the animal species in your quadrant before lunch?

....

2) Everbody must **submit** their practical books by 4pm.

....

3) They have **abandoned** their attempt to find a polymer as strong as steel.

....

4) If you **judge by** this thermometer then the temperature in this room is -20°C!

....

5) **Proceed with** the experiments that were suggested by Masters and Bromson.

....

6) Could you **explain** the background of your research in more detail?

....

7) Please **describe** the daily responsibilities of my new position as the lead gastroenterologist.

....

B) Rewrite the following sentences without changing the meaning. For example:

He is **bringing out** a new science textbook.
He is **publishing** a new science textbook.

1) How long will it take to **get through** the isolation and purification stages of this protein?

....

2) I will have to complete and **give in** the report on drug toxicity before Thursday.

....

3) We should **give up** this approach and try to find another way to synthesize this antibiotic.

....

4) **Going by** the reading on this gauge the pressure is 60 pounds per square inch.

....

5) If you **go by** the pole star, north is over there.

....

6) Now that we have the funding we can **go ahead with** the trip to the Arctic.

....

7) In the second half of the seminar he **went into a** description of the machine's components.

....

8) I must make an urgent call but the departmental telephone network has **gone down.**

....

9) He could tell by the silence that the lecture had not **gone down** well.

....

10) Every doctor in the UK must **get through** their membership exams before they can become a consultant.

....

UNIT 7: PHRASAL VERBS (GO - KEEP)

Go off (I): (Go, went, gone + off) To decay/deteriorate/spoil.
"The solution must be kept in the freezer otherwise it will **go off** very soon."
"If you leave a piece of meat outside, as it **goes off** it attracts insects."
"Fetal calf serum must be kept at -20°C for long term storage, otherwise it will **go off**."

Go on (N): (Go, went, gone + on) To continue.
"What you said about the assembly of the synchrotron is interesting; could you **go on** and explain this further?"
"The meeting **went on** for twenty minutes longer than expected."
"The weather forecaster said that the rain would **go on** all day."
"The specialist in Public Health Medicine has calculated that the incidence of tuberculosis will **go on** increasing and will double in the next ten years."

Go over something (N): (Go, went, gone + over) To repeat/to outline/to recapitulate.
"Could you **go over** your explanation of how this works again?"
"Now, I would just like to **go over** the important points that I have covered in this talk."
"Would g**oing over** what you will say help you to prepare for the television interview tomorrow?"
"The patient clearly didn't understand everything you told him. You will probably have **to go over** your description of his management the next time you see him."

Go (a)round (N): (Go, went, gone + (a)round) To spread/circulate.
"There is a rumour **going around** that the company is about to be taken over."
"There is not enough money in the National Institute of Health to **go around**."
"If you put a drop of dye in this flask, it will take 60 seconds **to go round** the whole container."

Go up (N): (Go, went, gone + up) To increase.
"The velocity of the object **went up** from 10 to 30 m/s."
"In the summer the temperature can **go up** by as much as 20°C."
"In the centre of the city the humidity **went up** dramatically."
"The prevalence of tuberculosis in New York **is going up** steadily."

Have something out (I): (Have/has, had, had + out) To publish.
"I must **have** a publication **out** before the end of the year."
"I wanted to **have** this paper **out** quickly but it has just been rejected by the referees."

Hunt down/up (N): To find/to search for and find.
"I'll be back after I have **hunted down** this reference."
"I have been trying to contact him all day, could you **hunt** him **down** and ask him to call me?"
"There is a reference to the British Medical Journal that I have found difficult to **hunt down**."

Keep to something (N): (Keep, kept, kept + to) To adhere to/to follow/to conform to.
"If we **keep to** the schedule we will finish the project on time."
"Let's try to **keep to** what is relevant to this discussion."
"We must **keep to** this chemotherapy regime when treating this group of patients."

Keep up with someone (N): (Keep, kept, kept + to) To try to be as good as /to follow/to stay with.
"If you want to excel in this field, Professor Kemp is the man to **keep up with**."
"When he starts talking I simply can't **keep up with** him; he is really clever."
"There are so many new publications in General Practice that it is difficult to **keep up with** the latest developments."

UNIT 7: PHRASAL VERBS - (GO – KEEP) EXERCISES

Go off	**Go on**	**Go over**	**Go (a)round**	**Go up**
Have out	**Hunt down**	**Keep to**	**Keep up with**	

A) Use one of the above words or phrases to rewrite the sentences below. For example:

No countries will **continue** producing chlorofluorocarbons after 2020.
No countries will **go on** producing chlorofluorocarbons after 2020.

1) If you store the reagent at room temperature it will **decay** in a week.

....

2) The company has decided not to **continue** with the development of laser weapons.

....

3) Please **outline** your strategy for developing a cure for AIDS.

....

4) There is a rumour **circulating** that you may win this year's Nobel prize.

....

5) The number of species in this area has **increased** by 15%; so increasing biodiversity.

....

6) By the end of the year I should have three more articles **published** in Nature.

....

7) Could you go back to the library and **search for** this reference?

....

8) We must **follow** our plan to build a robotic surgeon within five years.

....

B) Rewrite the following sentences without changing the meaning. For example:

Storing food in a fridge stops it from **going off** quickly.
Storing food in a fridge stops it from **decaying** quickly.

1) Is there gold in this soil or should I ignore the rumour **going around**?

....

2) The cost of the Scientific American has **gone up** by 10%.

....

3) We are planning to **have** the article **out** by March.

....

4) I had no difficulty **hunting down** the telephone number of the Science Research Council.

....

5) The department will be permanently closed if we do not **keep to** the safety regulations.

....

6) The Ferrari does not have enough power to **keep up** with the Lotus; the Lotus will win.

....

7) A literature search should help you **hunt down** the important article.

....

8) The number of people applying for fellowships in Medical Education has **gone up** by 35%.

....

9) Both the employer and employee must **keep to** the terms of the contract.

....

UNIT 8: PHRASAL VERBS (KEY - LOOK)

Key in(to) something (I): (1) To enter data into a computer/calculator etc. (2) To type in.
Key something in(to)

"Immediately after you **key in** all of the results, save the data."
"I wrote the program on a piece of paper and then **keyed** it **into** the desktop."
"Don't try to do the sum in your head, **key** it **into** the calculator."

Let something out (I): (Let, let, let + out)
(1) To allow to become public/to permit open knowledge.
(2) To release.

"We don't want to **let** these results **out** until we have confirmed them."
"At this setting the oxygen cylinder **lets** the gas **out** slowly into the incubator."
"The results of the interviews for the next Vice-President of Research and Development are being kept confidential until the successful candidate has been informed. We can't **let** the information **out** before then."

Look at (N): To examine.

"If we **look at** this diagram of the *population number* against *time* we can see that the species decreased in number rapidly after the predator was introduced."
"**Looking at** the specimen closely under the microscope allowed me to identify the genus and species."
"If you **look at** the patient's medical history more closely, you will see that this is her fourth admission to hospital this year. Each time she has been seeking the drug pethidine."
"If you **look at** the bullet under a microscope, you can see the specific striations that match this gun barrel."

Look forward to (N) To anticipate (with pleasure/optimism).

"I am **looking forward to** going to the meeting in New York; I'm going to take a holiday at the same time."
"He **looked forward** to buying up-to-date equipment for the new offices."
"The best thing about being promoted to Head of the Cardiology department is the increase in my salary."
"I **am looking forward** to my monthly paycheck."

Looking into (N): To investigate.

"If there is a suspicious sudden death, forensic scientists are usually asked to **look into** the cause of death."
"Chemists are **looking into** the possibility of making new organometallic compounds."
"We are **looking into** the possibility of doubling the number of laboratory staff in this department."

Look over (N): To review.

"I wrote this summary in a hurry late last night, could you **look over** it and make sure it's alright?"
"After she **looked over** the instructions again she decided to try to operate the machine."

Look up something (N): To search for a reference/item/fact.
Look something up

"I can't remember this organism's species; I'll **look it up** in an atlas later."
"**Look up** the specific heat capacity of the compound in the data book."
"**Look up** the atomic number in the periodic table."
"**I looked up** the precise dosage in the British National Formulary."

Look up to someone (N): To have in high regard/to admire/respect.

"I **look up to** any person who can design and build their own computer."
"The professor's secretary does not **look up** to him."
"I have **looked up** to him since we were at school; he was brilliant even then. Now he works for NASA."

UNIT 8: PHRASAL VERBS - (KEY – LOOK) EXERCISES

Key in(to)	**Let out**	**Look at**	**Look forward to**
Look into	**Look over**	**Look up**	**Look up to**

A) Use one of the above words or phrases to rewrite the sentences below. For example:

Do biologists **admire** physicists?
Do biologists **look up to** physicists?

1) To get money from the machine you must **enter** your personal identification number (PIN).

....

2) The car's exhaust **releases** toxic gases into the atmosphere.

....

3) If you **examine** the cell under the microscope you can see several nuclei.

....

4) We are **optimistically anticipating** the results of the drug tests on humans.

....

5) Maybe you should **investigate** the possibility of cloning this gene.

....

6) I gave the article to the professor **to review** before I submitted it to Nature.

....

7) I can never remember the exact speed of light, I always have **to search for** the precise value.

....

8) Dermatologists need to **examine** skin lesions carefully and describe them accurately.

....

B) Rewrite the following sentences without changing the meaning. For example:

If you **look at** this part of the spectrum you can see the peak caused by water.
If you **examine** this part of the spectrum you can see the peak caused by water.

1) I hope the development of voice activated computers means that I will not have to **key in** information.

....

2) If you turn on the tap you will **let** the water **out** at a pressure of 60 lbs per square inch.

....

3) If you **look at** the sun through a telescope, you may be blinded.

....

4) I look **forward to** the result of the organic synthesis; I am sure that the yield will be high.

....

5) I am **looking into** the possibility of buying a portable PC (personal computer).

....

6) After **looking over** his mathematical proof for the third time, I found his mistake.

....

7) **Look up** the hyperkalaemia protocol in the acute medicine handbook.

....

8) Because he has won the Nobel Prize for Chemistry, many people **look up to** him.

....

9) I am **looking forward to** a percentage yield of at least 80%.

....

10) Have you **looked into** the possibility that the disease is a vasculitis rather than a lymphoma?

....

UNIT 9: PHRASAL VERBS (MAKE - PUT)

Make up something (N): (Make, made, made + up) (1) To prepare. (2) To invent / To fabricate.
"I need to **make up** a 10 mM solution of sodium chloride; do you know where the powder is kept?"
"Please wait here while the pharmacist **makes up** your prescription."
"His reputation was destroyed when it was discovered that he had **made up** the results he had published."

Make something up to (N): (Make, made, made + up to) To increase to the required level.
Make up something to
"**Make up** the volume of the solution **to** 500mls."
"The foundation will give you some of the money but you will have to **make it up to** the full amount."
"There are too few women in this clinical trial, **to make up** the two halves **to** parity we must enroll 30 more women."

Make something out (I): (Make, made, made + up to) To see clearly/to discern/to distinguish.
Make out something
"You can just **make out** the two objects if the resolution of the microscope is good."
"If you look carefully at the spectrum you can just **make out** a peak that shows that the sample was contaminated."
"I have studied the ECG carefully and have not been able **to make out** an arrhythmia; the ECG looks normal."

Nod off (I): To fall asleep.
"Karen woke me up after I **nodded off** during the lecture."
"Nobody ever **nods off** while the director is speaking."
"I **nodded off** whilst the telephone call centre assistant described the limitations of the insurance policy.
"The patient **nodded off** before the anaesthetist had finished the intravenous injection."

Pick up (I): To understand/to comprehend.
"Whilst training as a biochemist, I **picked up** a little organic chemistry."
"I didn't **pick up** all the points she made but I was able to see the importance of her work."
"While reading the article I **picked up** the impression that this area of research might easily be investigated by our group."

Play up (I): To malfunction/to work incorrectly.
"I couldn't measure the potential difference; the voltmeter was **playing up**."
"A lot of electrical equipment starts to **play up** if the room temperature is raised."
"We cannot afford to have a respirator in the Intensive Care Unit that is **playing up**."

Pull out (of) (N): To withdraw from/to remove.
"We had to **pull out of** the conference because we would not have a presentation ready in time."
"It may be time to **pull out of** this project; it is unlikely to be successful."
"**Pulling out of** the collaboration became necessary after we discovered that their studies were not reliable."
"We have **pulled out of** the project to build a new general hospital but are instead planning to build an extension to this hospital."

Put something down to (N): (Put, put, put + down to) To attribute to/to assign to/to be caused by.
"He **put** these surprising data **down to** imperfect technique."
"I **put** the spelling mistakes **down to** the fact that I was using an unfamiliar keyboard."
"The meteorologist **put** the warm weather **down to** the high pressure and lack of wind or rain."
"Although the patient's recovery could be **put down to** a miracle, I think it is more likely that the cancer was successfully removed during surgery."

UNIT 9: PHRASAL VERBS - (MAKE – PUT) EXERCISES

Make up	**Make up to**	**Make out**	**Nod off**
Pick up	**Play up**	**Pull out (of)**	**Put down to**

A) Use one of the above words or phrases to rewrite the sentences below. For example:

In the middle of the toxicology conference I **fell asleep**.
In the middle of the toxicology conference I **nodded off**.

1) **Prepare** a 5% solution of $NaHCO_3$ in water.

....

2) If the telescope's resolution is not good you will not be able **to see** adjacent stars.

....

3) I never **understand** everything in a discussion about genetic technology.

....

4) The pharmacist will take 10 minutes **to prepare** your medicine.

....

5) Take 2 mls of the supernatant and **increase** it to 5 mls with water.

....

6) Good scientific communication will prevent people in the audience from **going to sleep**!

....

7) The government **assigned** the increase in cancer deaths to poor nutrition.

....

8) The weighing scale must be **malfunctioning** again, I don't believe its readings.

....

9) His success in academic medicine was **attributed to** his high IQ.

....

B) Rewrite the following sentences without changing the meaning. For example:

The plane fell 2,000 feet but fortunately **pulled out of** the dive 100 feet from the ground.
The plane fell 2,000 feet but fortunately **withdrew from** the dive 100 feet from the ground.

1) To make a threefold dilution, take 1ml of solution and **make it up** to 3 mls with the solvent.

....

2) If you look into the distance you can **make out** one of our planes.

....

3) Because I was up carrying out an experiment late last night, I **nodded off** during today's physics lecture.

....

4) Could you repeat the explanation, I did not **pick up** everything you said.

....

5) If the centrifuge starts to **play up** call the repairman.

....

6) This satellite project will fail; it's time to **pull out** and leave it to the Americans.

....

7) The investigators **put** the plane crash **down** to a mistake by the pilot.

....

8) This patient has three first degree relatives who have had myocardial infarctions under the age of 60 – **we put this down to** a family history of hypercholesterolaemia.

....

9) The spectrometer **is playing up** again. I think it is time to buy a new one.

....

UNIT 10: PHRASAL VERBS (PUT - REEL)

Put in for (something) (I): (put, put, put + in for) To apply for.
Put in something for

"I **put in for** the grant and was confident of being awarded at least £100,000."
"I am recommending that my student **puts in for** a scholarship - I will be on the interviewing committee."
"This is such a good piece of research that we should be able to **put it in for** an award."
"After having my IQ tested **I put in for** membership of MENSA."

Put something off (I): (put, put, put + in for) To delay/postpone.
Put off something

"He usually **puts off** writing his articles until the very last moment."
"It is not possible to **put off** making a decision any longer."
"**We put off** starting the ward round until the consultant and the nurse arrived."

Put on (N): (put, put, put + on) (1) To switch on/to turn on. (2) To wear clothes/clothing.

"**Put on** the ultraviolet light to sterilize the room after you leave."
"**Put on** the protective clothing before you enter the room."
"In an effort to maintain an aseptic environment you must take off your jacket and **put on** a white coat and gloves."

Put out (N): (put, put, put + out) To extinguish/turn off (*often a light*).

"He **put out** the flame and turned off the gas."
"The last thing that he did before he left the building, was to **put out** the lights."

Put someone through to (N): (put, put, put + through to) To connect to/to put in contact with.

"Could you **put me through to** the assistant professor?"
"The operator **put me through to** the director's secretary, after about 10 minutes."
"I phoned the British Medical Association main offices but **was put through to** the BMA helpline."

Read something into something (I): (read, read, read) To interpret/to find hidden messages or meanings.

"You cannot **read** too much **into** these results; you must carry out more experiments to support your inferences."
"I could be **reading** too much **into** what the group leader said, but I believe he really wants to give Dr Bright the job not Dr Shore."
"You should **not read anything into** the fact that he has been offered membership of the Royal College of Science and you haven't."

Read up on something (N): (read, read, read up + on) To review written material/to study.
Read something up
Read up something

"My molecular biology is not very good, I really need to **read up on** it."
"I simply do not have the time to **read up** as much as I would like in this field."
"I have **to read up on** the latest guidelines for the management of asthma."

Reel something off (I): To speak/repeat *(quickly and easily).*
Reel off something

"He **reeled off** the statistics so rapidly that it was clear he had memorized them."
"He **reeled** the statistics **off** so quickly that it was clear he had memorized them."
"Tom learned the first twenty elements of the periodic table and then **reeled** them **off** in front of us."
"She **reeled off** the key findings in the history and examination and offered a confident diagnosis."

UNIT 10: PHRASAL VERBS - (PUT - REEL) EXERCISES

Put in for   Put off   Put on   Put out

Put through to   Read into   Read up on   Reel off

A) Use one of the above words or phrases to rewrite the sentences below. For example:

I decided to **delay** buying an Apple MacIntosh or an IBM PC until we knew our needs.
I decided to **put off** buying an Apple MacIntosh or an IBM PC until we knew our needs.

1) I have **applied for** a grant from the National Institute of Health.

....

2) The government **delayed** making a decision about the drug until new safety tests were done.

....

3) **Wear** your lab coat!

....

4) **Turn off** the ultraviolet light before you enter the room.

....

5) Operator, could you **connect** me **to** the head office of BRITCO Pharmaceuticals?

....

6) Many people believe that we are **overinterpreting** the loss of the ozone layer.

....

7) Because she had **studied** protein purification techniques, she prepared an excellent protocol.

....

8) Her complete understanding of her field allowed her to **repeat** facts confidently.

....

9) The statistician does not believe that our data is significant at the $P<0.05$ level; it looks as though we were **overinterpreting** the data.

....

B) Rewrite the following sentences without changing the meaning. For example:

I want to **read up on** the new synthetic materials - they may be useful in our research.
I want to **review** the new synthetic materials - they may be useful in our research.
It is common for inexperienced researchers to **read too much into** their data.
It is common for inexperienced researchers to **overinterpret** their data.

1) I have **put in for** the job of Director of Research and Development.

....

2) I shall **put off** applying for a job as a geologist until I have got my doctorate.

....

3) The streetlights are **put out** automatically at dawn each day.

....

4) If Professor Bell's secretary does not **put** me **through to** him, I can contact him by e-mail.

....

5) He told me that I was **reading** too much **into** my data - I needed to do more experiments.

....

6) The salesman confidently **reeled off** the specifications of the music system.

....

7) **Put** the colorimeter **on** 30 minutes before you use it.

....

8) Having **read up on** the physics of subatomic particles, I was able to make far more intelligent suggestions during my conversation with him.

....

9) After **reading up on** the pathology of liver diseases, I was able to learn far more during my attachment in medical gastroenterology.

....

UNIT 11: PHRASAL VERBS (ROUND - RUN)

Round something off (N): To abbreviate/finish.
Round off something

"You ought to **round** those numbers **off** to three significant figures."
"You ought to **round off** those numbers to three significant figures."
"Finally, the meeting was **rounded off** by agreeing a date for the next meeting."
"He **rounded off** his lecture by referring to all of the laboratory's previous successes."

Rub off (on) someone (I): To be transferred to/to be passed on to.

"If he were to join the group his enthusiasm might **rub off on** the others and their work would improve."
"Genius is not something that usually **rubs off** from one individual **on**to another."
"Because the bacteria are transmitted by contact, it is **rubbed off on**to individuals who touch each other."

Rule someone/something out (N): To exclude.
Rule out someone/something

"We cannot **rule out** random error as a cause for these anomalous results."
"His application was **ruled out** after it was discovered that he had a criminal record."
"The lack of abdominal pain **rules out** appendicitis."

Run across someone/something (N): (Run, ran, run + across) To meet by accident/To see by chance.

"I was very surprised to **run across** Dr Sims at the Milan conference because I thought he was in the US."
"If you **run across** any relevant facts while you are searching through the New England Journal of Medicine, please tell me as soon as possible."
"I was reading the Scientific American when I **ran across** the advertisement for a postdoctoral scientist."

Run at (I): (Run, ran, run + at) To function at/to operate at.

"The engine is most efficient when **run at** less than 1200 revolutions per second."
"The centrifuge must never be **run at** more than 20g."

Run something off (N): (Run, ran, run + off) To make/produce *(multiple copies)*.
Run off something

"Could you **run off** half a dozen copies of this article for me please?"
"I asked the printer to **run off** 2000 booklets not 20,000 booklets!"
"I asked the publisher to **run off** fifty more copies of my first book, for my friends and family."

Run out of something (N): (Run, ran, run + out of) To use up/to completely deplete.

"I think that there isn't any more ethanol in the lab; we have **run out**."
"It should be somebody's responsibility to make sure that we never **run out of** essential material."
"We have **run out of** paracetamol on the ward, so I will get some from the pharmacy."

Run over/through something (N): (Run, ran, run + over) To review/to go over.

"Could you **run over** the protocol with me?"
"Could you **run through** the protocol with me?"
"Before he attends an important meeting, he always **runs through** what he will say or do."
"After **running over** my Surface Anatomy speech three times, I was sure that it would go well."

UNIT 11: PHRASAL VERBS - (ROUND - RUN) EXERCISES

round off	**rubbed off**	**rule out**	**ran across**
run at	**run off**	**run out of**	**run over/through**

A) Use one of the above words or phrases to rewrite the sentences below. For example:

The doctors' concern about bioethics has been **transferred to** the scientists.
The doctors' concern about bioethics has **rubbed off** onto the scientists.

1) I **met** the Safety Officer in the library; he reminded me of the new regulations.

....

2) The DNA fingerprint evidence was enough to **exclude** him as a suspect.

....

3) He **finished** the article with a section on possible future technological developments.

....

4) **Operate** the chart recorder at 12 cm/min to get a satisfactory trace (record).

....

5) Could you **make** 7 copies of this diagram for me?

....

6) We've **used up the** butane, so the gas cylinder is empty.

....

7) There are one or two things I am not clear about, can we **go over** them again?

....

8) Can **we review** the risk factors for atherosclerosis?

....

9) The most recent experimental data **exclude** your favoured hypothesis.

....

10) It would be helpful if you could **go over** your proposed theory again. I didn't understand it the first time.

....

B) Rewrite the following sentences without changing the meaning. For example:

If you **look at** the cell under the microscope you can see the cell wall.
If you **examine** the cell under the microscope you can see the cell wall.

1) She **rounded off** the microbiological conference with a summary of recent discoveries.

....

2) We cannot **rule out** the possibility of electric polymers being used to make new TV screens.

....

3) It was during the Marine Biology Conference that she **ran across** her previous supervisor.

....

4) If you **run** the centrifuge **at** 2000 revolutions per minute you will separate the components.

....

5) The professor's secretary **ran off** 5 copies of the grant application.

....

6) I have **run out** of space on the hard disk.

....

7) Let's **run through** this report before the meeting starts.

....

UNIT 12: PHRASAL VERBS (RUN - SLIP)

Run something up (N): (Run, ran, run + up) (1) To accumulate. (2) To produce/make.
Run up something
"The spending in the Radiology department is out of control; we have **run up** an enormous electricity bill this month."
"I'll try to **run up** a poster in time for the presentation on the 1st of March."
"How long will it take to **run up** a batch of this new electric plastic?"

Screw something up* (I): To make a mess/to damage or destroy. (***Used as an insult.***)
Screw up something
"This piece of equipment was working fine until you touched it; you have completely **screwed** it **up**."
"After **screwing up** the experiment for the third time, I asked the laboratory technician for some advice."
"If you **screw up** then your errors will be discovered at the peer review stage."

Send down (N): (Send, sent, sent + down) To be expelled from university (U.K.).
"His career has been held back because he was **sent down** from Oxford University."
"Do you know when he was **sent down** from Cambridge University Medical School?"

Send (off) for (N): (Send, sent, sent + (off) for) To order by post/To write off for.
"I have **sent off for** the necessary equipment; it should all arrive within 14 days."
"We **sent off for** the drug six weeks ago but it has not arrived yet."

Set someone/something back (N): (Set, set, set + back) To impede/to hold back/prevent.
Set back someone/something
"His career was **set back** after he had to withdraw a published article."
"His financial adventure has **set** us **back** by £1,000."
"The malpractice legal action against Dr Good was successful. The financial settlement the hospital must make will **set us back** two million pounds."

Set up something (N): (Set, set, set + up) (1) To start/to establish. (2) To assemble/to organize.
Set something up
"The professor of surgery wants to **set up** a new research project."
"When the sales representative **set up** the new equipment it all worked perfectly but it broke down immediately after he left."
"The support staff **will set up** this computer with the appropriate settings for this department, then you will be able to use it."

Settle down (N): To calm down.
"We must wait for the wind to **settle down** before we can go outside and take measurements."
"After the signal on the cathode ray oscilloscope **settled down**, it was possible to see that the signal consisted of at least two waves."
"After his initial comments the audience was so angry that he had to wait five minutes for them **to settle down**."

Show up (N): (1) To appear/to be present. (2) To be visible.
"I'm not sure that many people will **show up** for the meeting."
"Only five botanists **showed up** at the ecological conference."
"The aircraft will **show up** on radar before you see it with the naked eye."
"The junior doctor **showed up** late for his first day at work."

Slip up (N): To make a mistake.
"I know I have **slipped up** but I can't understand how."
"The secretary has **slipped up** again; she has written **herring** instead of **heroin**."

UNIT 12: PHRASAL VERBS - (RUN - SLIP) EXERCISES

Run up	**Screw up**	**Send down**	**Send off for**	
Set back	**Set up**	**Settle down**	**Show up**	**Slip up**

A) Use one of the above words or phrases to rewrite the sentences below. For example:

Make a sample of your aluminium and lithium alloy for strength testing.
Run up a sample of your aluminium and lithium alloy for strength testing.

1) He **wrecked** the computer after accidentally pouring coffee on the keyboard.

....

2) He was **expelled** from university for stealing drugs from the pharmacology department.

....

3) I can't buy a Geiger counter here, I'll have to **order** one.

....

4) That surprising result **held back** our work for 8 months.

....

5) Five years ago the physics department **started** a project to develop an electronic nose.

....

6) The animal can change colour so that it **is** not **visible** against the background.

....

7) I am less likely to **make a mistake** using a Windows based program than an MSDOS based program.

....

8) All flights from the airport are cancelled until the weather **calms** down.

....

9) The junior doctor discovered that his job applications were being **impeded** by a poor reference.

....

10) I have just **written off for** the latest edition of Gray's Anatomy.

....

11) After taking the antacid tablets, her abdominal pain **calmed down**.

....

B) Rewrite the following sentences without changing the meaning. For example:

There is a rumour that he is **setting up** a clinical trial for the treatment of AIDS.
There is a rumour that he is **organising** a clinical trial for the treatment of AIDS.

1) It will take 2 weeks to **run up** a miniature model of the plane and then test it.

....

2) Cigarettes **screw up** your lungs and circulatory system.

....

3) Is it better to get a bad degree classification than to be **sent down**?

....

4) After seeing the advertisement for a toxicologist I **sent off** for an application form.

....

5) Losing his data would have **set** him **back** for months but fortunately he had backed up a copy of the data.

....

6) The computer network is **set up** with a password system to protect the user's data.

....

7) It will take 4 hours to **set up** the experiment but only 30 minutes to perform it.

....

8) The cardiology consultant **slipped up** and missed a cardiac murmur.

....

UNIT 13: PHRASAL VERBS (SLOW - TAKE)

Slow down (N): To decelerate/to decrease in velocity or speed.
"The horse **slowed down** as it went up the hill."
"The plane is **slowing down** to land."
"The patient has a tachycardia, so we must **slow down** his heart."

Speed up (N): To accelerate/to increase in velocity or speed.
"The car **speeded up** from 20 to 60 mph."
"Could you **speed up**, I am late for an important appointment."
"**Speed up** the intravenous delivery of morphine via the infusion pump."

Spell something out (N): To make something clear/to clarify.
Spell out something
"I did not completely understand what you just said, could you **spell** it **out** to me."
"When you are teaching students you must be prepared to **spell out** your message."
"It is not clear from the patient history and examination, why that particular antihypertensive drug was chosen. Could you ask the prescribing doctor to **spell out** his reasoning?"

Stand out (N): (Stand, stood, stood + out) To be prominent/to be obvious.
"The C=C bond peak **stood out** particularly in that infra red spectrum."
"Einstein **stands out** as one of the greatest scientists of the 20th century.
"The thing that always **stands out** as you listen to him speak, is his great enthusiasm."
"The symptom that **stands out** is the symmetrical erythematous rash. This is the key to the diagnosis."

Sum up (N): To summarize/to conclude.
"You have been talking for a while, could you **sum up** now?"
"At the end of the report the positive features of the product were **summed up**."
"During the ward round, the key findings for each patient were **summed up**."

Switch off (I) (1) To turn off. (2) To lose concentration.
"**Switch** the lights **off** if you are the last to leave."
"The talk was so boring that I **switched off** after 10 minutes and I started to think about what I would do during my holiday."
"Whenever anybody starts talking about disease pathogenesis, I **switch off**."

Take something down (N): (Take, took, taken + down) To write down a communication.
Take down something
"Could you **take down** this message and send a copy to the Mathematics, Physics and Engineering departments."
"I can never **take down** notes quickly enough during these seminars."

Take something in (N): (Take, took, taken + in) To absorb.
Take in something
"She **takes in** information very quickly and is then able to use it effectively."
"His accent sometimes makes it difficult to **take in** what he says immediately."
"Studies have shown that patients only **take in** 10% of what General Practitioners tell them."
"Sponges are effective because they can **take in** the equivalent to their own volume of water."

Take off (I): (take, took, taken + off) (1) To be successful/popular. (2) To remove/subtract. (3) To become airborne (planes).
"Her career **took off** after she left the university and started a biotechnology company."
"After **taking off** the discounts from the original selling price, the laser printer became a good buy."
"The plane has to be moving at 80 mph before it **takes off**."

UNIT 13: PHRASAL VERBS - (SLOW - TAKE) EXERCISES

Slow down	**Speed up**	**Spell out**	**Stand out**	
Sum up	**Switch off**	**Take down**	**Take in**	**Take off**

A) Use one of the above words or phrases to rewrite the sentences below. For example:

If we equip this laboratory with the best equipment we will **accumulate** a huge bill.
If we equip this laboratory with the best equipment we will **run up** a huge bill.

1) You must **decrease your speed of** speech for the voice recognition system to work.

....

2) Could you **clarify** your plans for your research division over the next 5 years?

....

3) The water peak **is prominent** in this spectrum.

....

4) He **summarized** his model of how the brain works.

....

5) As he described the Schrödinger equation, most of the students **stopped concentrating**!

....

6) If the speaker talks too quickly you will not have time to **write down** everything.

....

7) The latest modem **became popular** last summer when sales increased dramatically.

....

8) The aspect of the presentation that **is most prominent**, is the discussion of cancer prevention by dietary alterations.

....

B) Rewrite the following sentences without changing the meaning. For example:

Nonylphenols are acting to **set back** the development of the male reproductive system.
Nonylphenols are acting to **impede** the development of the male reproductive system.

1) He speaks so clearly that it is very easy to **take in** what he has said.

....

2) The electric field **speeds up** the electrons by 500 m/s.

....

3) The programming manual would be better if it **spelt out** the instructions.

....

4) The text **stands out better** on a light emitting diode display than on a liquid crystal display.

....

5) The science reporter **took down** every word that the Professor of Genetics said.

....

6) You must exit the program before **switching off** the computer.

....

7) The rocket **slowed down** to 5 m/s before it landed on Pluto.

....

8) He **summed up** the multimedia potential of the latest personal computer.

....

9) You have to **switch off** all mobile phones before entering the hospital; this is particularly important for the Intensive Care Unit.

....

UNIT 14: PHRASAL VERBS (TAKE - WADE)

Take on (N): (Take, took, taken + on) To agree to do/to handle.
"This contract for us to supply Apple McIntosh is too big for us to **take on**."
"Are you sure that you can **take on** such a huge amount of work? You must not risk your health."
"**Taking on** this project means carrying out the largest randomized controlled trial in the country."

Turn down (N): (1) To reject. (2) To decrease intensity.
"My application for that job was **turned down**; I'll try again next year."
"The sound is too loud; **turn down** the speaker."
"I **turned down** the screen brightness."
"The application to appointment ratio for this Anaesthetics Registrar job is 9:1 so we will have to **turn down** eight of the candidates."

Turn something off (N) To switch off.
Turn off something
"The plane landed and then its engines were **turned off**."
"**Turn off** the radio and do some work!"
"It became clear that the patient was brain dead and was only being kept alive by the respirator; somebody would have to make the decision to **turn off** that machine."

Turn on (N): To switch on.
"The central heating system had not been **turned on**."
"If I **turn on** the ultraviolet light and leave it on overnight, the incubator will be sterile by the morning."
"Please **turn on** the plasma screen."
"The Nuclear Magnetic Resonance Spectrometer must be permanently **turned on**."

Turn out (I): (1) To result/To be the result. (2) To conclude.
"How did the experiment **turn out**? Did you get the results you wanted?"
"The presentation **turned out** much better than expected."

Turn to (N): To look towards/to refer to.
"**Turning to** figure 5a we can see that the number of species in this region has decreased steadily over the last 20 years."
"I think that we will have to **turn to** someone else to referee this paper."
"We should only **turn to** private sources of funding when all other possible sources have been tried."
"Whenever I want a definitive answer on any question in respiratory medicine, I **turn to** Dr Wheeze."

Turn up something (N): (1) To increase intensity. (2) To attend. (3) To arrive at.
Turn something up
"Please **turn up** the brightness on the television."
"Only 50% of the group **turned up** at the haematology conference."
"I **turned up** the gain on the cathode ray oscilloscope to demonstrate the waveform more easily."

Use up (N): To deplete/to consume all of something.
"There is no glucose left; it has all been **used up**."
"We have to share the memory on the hard disk equally between all of the staff, otherwise the computer's memory will be **used up** by a small number of individuals."

Wade through something (N): To get (or move) through slowly.
"I have a large spreadsheet **to wade through** before tomorrow; I am going to be up most of the night."
"Instead of **wading through** all of the patient's history why don't you give me a brief verbal summary?"

UNIT 14: PHRASAL VERBS - (TAKE - WADE) EXERCISES

Take on	**Turn down**	**Turn on**	**Turn out**	**Turn off**
Turn to	**Turn up**	**Use up**	**Wade through**	

A) Use one of the above words or phrases to rewrite the sentences below. For example:

We need someone to **handle** this research project.
We need someone to **take on** this research.

1) Put the glass equipment on the shelf and **switch on** the autoclave.

....

2) I **increased** the velocity of the centrifuge to 3000 rpm (revolutions per minute).

....

3) The thermostat will **decrease** the heating if the house becomes too warm.

....

4) Virtual reality **resulted in** more applications than first anticipated.

....to have more applications than first anticipated.

5) We will **handle** the task of designing and building the next Boeing plane.

....

6) I **referred to** my supervisor for technical advice.

....

7) **Switch on** the High Performance Liquid Chromatography machine and load the sample.

....

8) This hospital **is handling** more patients than it can manage; we need to start redirecting patients to nearby hospitals.

....

B) Rewrite the following sentences without changing the meaning. For example:

A thought experiment is always easier to **set up** than the real thing.
A thought experiment is always easier to **organise** than the real thing.

1) We calculate that the petroleum in this oil field will **be used up** in 2 years.

....

2) As the reagent (reactant) is used up the chemical reaction **slows down**.

....

3) It took me 6 hours to **wade through** the review article on the management of asthma.

....

4) I **turned up** at the conference just in time to give my presentation.

....

5) **Turn on** the Blu-ray player and play the Region 1 disc.

....

6) I **turned to** the Annual Review of Physics for more information.

....

7) The printing press runs continuously; it is not **turned off**.

....

8) The lights were **turned down**, the projector was switched on and the film was shown.

....

UNIT 15: PHRASAL VERBS (WARM - WRITE)

Warm up (N): (1) To initiate before proper functioning. (2) To heat up/to make warm.
"You have to leave the photocopier on for a few minutes to **warm up** before you can use it."
"Does the printer work as soon as you turn on the power or does it need a little time to **warm up**?"
"The salt dissolves faster if you **warm up** the solution a little."
"The microscope comes on immediately, you do not have to wait for it to **warm up**."
"I was waiting for the dialysis machine to **warm up** when my bleep went off and I had to go to the rehabilitation ward."
"For a defibrillator to be useful it must **warm up** in less than a second."

Weigh something up (N): To assess/consider.
Weigh up something
"We must **weigh up** the advantages and disadvantages of each plan before reaching a conclusion."
"After **weighing up** the strengths and weaknesses of each candidate, I have decided to offer the postdoctoral position to Dr Travers."
"I **weighed up** the location, salary and prestige of working in Cambridge against working in Oxford."
"In choosing a medical speciality you should **weigh up** your career aspirations, lifestyle, colleagues and likely remuneration, together with your aptitude for the speciality."

Work something out (N): To deduce/to calculate/to reason.
Work out something
"I need to think further about the problem so that I can **work out** the best solution."
"A great chess player must consider several options before he **works out** what to play."
"You can **work out** the prognosis for a woman with breast cancer by using the Nottingham Prognostic Index."

Write something up (N): (Write, wrote, written + up) To write an account of something.
Write up something
"I have been offered a job so I must have my thesis **written up** by the autumn."
"I **wrote up** the final draft (=version) of the article last night."
"The review article on the efficacy of statins in the management of atherosclerotic disease has to **be written up** by Thursday the 6th of May."
"A summary of all the work carried out by the Scientific Services Division in the last two years must be **written up** and delivered to the Head Office by Friday."

UNIT 15: PHRASAL VERBS - (WARM - WRITE) EXERCISES

Warm up	Weigh up	Work out	Write up

A) Use one of the above words or phrases to rewrite the sentences below. For example:

He **worked out** that there were 10000 possible combinations for the password.
He **calculated** that there were 10000 possible combinations for the password.

1) The tyres on the car **become warm** as the car speeds up.

....

2) A Japanese company has **deduced** a way to make a computer the size of a pea.

....

3) Organic chemists have **found** a way to synthesize the sugar.

....

4) I **considered** the alternatives before making a decision on where to drill for petroleum.

....

5) It does not take long to **write** an abstract.

....

6) She **assessed** the advantages and disadvantages of each antipsychotic medication.

....

7) As the metal **heats up,** its electrical properties change.

....

8) I **calculated** the probability of rain as 70%.

....

9) I **calculated** the required dosage by allowing 1mg for each 10kg of bodyweight.

....

B) Rewrite the following sentences without changing the meaning. For example:

I can't **work out** how to use the equipment.
I can't **deduce** how to use the equipment.

1) After you **turn** the machine **on** it must warm up for 2 minutes before you can use it.

....

2) **Work out** a solution to the mathematical equation.

....

3) I **weighed up** the consequences of staying there against starting my own software business.

....

4) Someone has **worked out** that the mercury in tooth fillings causes kidney damage.

....

5) As part of the grant application he had to **write up** a report of the group's work.

....

6) The oven must **warm up** before it can be used.

....

7) The journalist **wrote up** an account of his interview with Professor Crick.

....

8) He has **worked out** how to synthesize a reagent more versatile than the Grignard Reagent.

....

UNIT 16: CONNECTIVES - ADDITION

The use of connectives in technical and scientific English is important in creating **logical arguments**. It is **good style to vary the use** of connective words that have similar meanings. This unit describes common connectives that have the meaning of *addition*.

And (N) = In addition. It is used to link two or more items that support each other; it is rarely used at the beginning of a sentence:
"Sodium hydroxide **and** sodium hydrogen carbonate are both water soluble."
"I would like to order 10 blu-ray discs **and** 15 compact discs."

In a list, each of the items is separated by a comma. **And** is used between the last two items:
"The greenhouse effect**,** the hole in the ozone layer**,** acid rain **and** the decrease in the area of the rain forest are issues of international concern."
"Kinetic theory assumes that the molecules have almost no volume, spend negligible time in collision, move with almost uniform velocity **and** experience nearly no intermolecular forces."

In addition (N) = And. **In addition** is frequently used *at the beginning* of a statement:
"I think that we have all the materials that we need to do this experiment. **In addition,** both Richard and Yvonne will be free to do it tomorrow."
"**In addition** to attending the French and Italian conferences, we could have a small holiday."

Also (N) = In addition. **Also** cannot be placed at the end of the clause:
"The cost of the apparatus is not the most important factor to be considered in buying a spectrometer; we must **also** consider the availability of maintenance and repair support."
"The effect of temperature and pressure on the aging of the polymer have been considered, we must **also** consider the effects of radiation."
"**Also**, I think we should give this area of research more time and money; there is a real possibility that it may lead somewhere."

American English allows the use of also **at the end of the phrase:**
"If I have time I will attend the seminar also.**"**

Furthermore (F) = Additionally/moreover. This word is used in more formal contexts. Furthermore is used at the beginning of the phrase:
"There is no suggestion of how it might work nor has the treatment been evaluated clinically; **furthermore,** there have been recent unofficial reports that the drug has serious side effects."

Moreover (F) = Furthermore. It is used in more formal English (e.g. in a presentation) to introduce a new statement or concept. Sometimes the introduced statement is something that was almost forgotten (an afterthought); and the level of emphasis may be higher/lower than the earlier statements:
"The bottle of liquid has no label, no date and has not been claimed by anyone. **Moreover** the new safety regulations require all substances to be registered; I think we must dispose of the bottle carefully."
"His research seems to be producing few results. **Moreover** I do not think that we should introduce a person with such poor interpersonal skills into this group."

(I) = ***Informal***
(N) = ***Neutral***
(F) = ***Formal***

UNIT 16: CONNECTIVES - ADDITION (EXERCISES)

And	**Also**	**In addition**
Moreover	**Furthermore**	

A) Complete the following sentences with the most appropriate connective:

1) The organic compounds we will discuss in this lecture include alkanes, alkenes, alkynes ________ amides.
2) We have discussed the amount of fuel, the weight of the passengers and the distance to be traveled. However, we _______ need to consider the weight of the vehicle.
3) ________ to the surgical disciplines, pathology is also underrepresented in the management of this hospital.
4) You must note the kingdom, phylum, class, order, family, genus __________ species.
5) The known risk factors for urinary tract infections are gender, age, use of instrumentation and immunosuppression. For the first time, we have provided evidence that socioeconomic status is an independent risk factor. _________ this conclusion is supported by evidence provided by the NIH and NHS.
6) The physical properties in which we are interested are thermal conductivity, tensile strength _________ elasticity.
7) *Neutral scenario:*
It is wrong to prescribe medication to a patient without taking a history and carrying out an examination. It can be extremely dangerous for the patient. _________ it can lead to significant legal difficulties for the doctor and the hospital.
8) *Formal scenario:*
Men and women have different chromosome complements. _________ these differences lead to different phenotypes.
9) To calculate the gravitational force of attraction between two bodies, _________ to the mass of each body, the gravitational constant and the distance between each body are required.
10) This factor is _________ needed to assess the likely success of the project.
11) The total capacity of this building can be calculated by adding the number of rooms in the first, second, third ________ fourth floors.

B) Change the *italicized* connective in the following sentences to one with a similar meaning:

1) Children with Down's syndrome have very specific physical abnormalities, which include short stature and congenital cardiac disorders. The management of such problems can be very difficult. *Moreover,* their management can be made even more difficult by the learning difficulty that such children usually possess.

....

....

2) Intelligence (measured by IQ) is correlated with height, speed of information processing and academic success. *In addition*, recent research has shown a negative correlation with conscientiousness.

....

....

3) The halogens are fluorine, chlorine, bromine, iodine and astatine. *Also*, there is an "undiscovered element" called ununseptium that is believed to be a halogen.

....

....

4) The car crash was caused by a combination of driver error, mechanical failure and excessive speed. *Furthermore*, the effect of the collision was worsened by the design of the cars.

....

....

5) Scientific weather forecasting depends on measurements of temperature, humidity, atmospheric pressure, wind velocity and precipitation. *Moreover*, in recent years, computer modeling has become an essential weapon in the armory of a good meteorologist.

....

....

UNIT 17: CONNECTIVES - DEDUCTION

It is often necessary to reason with words when working in a technical field. For example: "We are not getting a picture on the screen, **so either** the transmitter is off **or** the receiver is broken."

Some important connectives used in logical reasoning are described below. In writing, it is generally poor style to repeat the same connective; it is better to know a range of connectives with similar meanings.

Thus (F) = As a result, in this way, so. Because "thus" is a formal word it is often used in written language:
"**Thus** you can calculate that the modem will take 10 seconds to send all of the information."
"**Thus** the fingerprint evidence and the dental records show that the dead man was Mr Jones."

Therefore (N) = Thus, accordingly, for that reason. "Therefore" is not as formal as "thus":
"It is a gaseous halogen, **therefore** it is either fluorine or chlorine."
"**Therefore**, the probability of throwing heads or tails with that coin, is the same."
"**Therefore,** with increasing pollution, the quality of food will decrease."
"The virus has a two week incubation, **therefore** she must have been infected whilst on holiday."

Hence (F) = Thus. "**Hence**" can be used to replace "**thus.**"
"**Hence** during the summer, when there is most pollen in the air, many employees cannot breathe easily."
"**Hence** the wind resistance, W, will be proportional to the speed, S, of the car."

So (I) = Therefore. "So" is most commonly used in speech.
"He isn't in the laboratory, the office or the toilet **so** he must have gone home."
"AIDS damages T lymphocytes, consequently damaging the immune system and **so** allowing secondary illnesses to occur."
"**So**, because hydrogen is lighter than nitrogen, the balloon rises in the air."

> Thus, therefore, hence **and** so **can be used at the beginning of a phrase to introduce a conclusion.**

Or (N) This conjunction is used to describe or discuss two or more alternatives.
"I know that she is either a zoologist **or** a botanist because she works in the biology department."
"The answer is either to increase the current **or** decrease the resistance of the electrical circuit."

Either.....or (N) This is used to discuss or describe two or more alternatives.
"To increase the maximum speed of the plane, you must **either** decrease its weight **or** decrease its wind resistance."
"It is a gaseous halogen therefore it is **either** fluorine **or** chlorine."
"To increase income we must **either** decrease costs **or** increase sales."
"Our investigation of the patient's pyrexia of unknown origin, suggests that the underlying cause is **either** a lymphoma **or** a vasculitis."

Neither......nor (N) This is used to exclude two or more alternatives.
"He could see **neither** blue **nor** green because he was colour-blind."
"**Neither** changing the temperature **nor** the volume **nor** the pressure of the gas will affect its mass."
"**Neither** of the methods that you have suggested is practical."

(I) = *Informal*
(N) = *Neutral*
(F) = *Formal*

UNIT 17: CONNECTIVES - DEDUCTION (EXERCISES)

For each of the scenarios below, complete the sentences with the appropriate connective(s) from this unit:

Scenario 1) A teacher talking to a student alone after class.
Teacher: "We know that the molecular weight of the compound is 28. We also know that the compound has an empirical formula of CH_2. Finally, we know that the compound is only composed of carbon and hydrogen. _________ the compound must be ethene. Do you understand?"
Student: "No."

Scenario 2) Extract of a discussion from a published article.
Small animals have a relatively high surface area. This means that they have a high surface area to volume ratio. ________ they are more susceptible to heat loss than large animals. Because of this they need to maintain a higher basal metabolic rate. Accordingly, small animals need to eat more food.

Scenario 3) A lecturer in nutrition talking to a general practitioner.
Nutritionist: There are essentially two alternatives if an individual has a body mass index of 33; ________ they are obese ________ they are very muscular athletes. In addition, you know that the patient is 5'11" tall and has a waist circumference of 40 inches. Which possibility do you favour?
General Practitioner: If these were the two alternatives then I would favour the conclusion that the individual is obese. However I think your scenario is very simplistic, for example, couldn't he be a fat athlete?

Scenario 4) Two students trying to understand some physics during their lunch break.
Student 1: The force on an object is a product of its mass and the acceleration of that object. Considering the weight of an object in similar terms, then weight is a product of its mass and gravity. ________ if the mass of the object is constant and gravity is constant, then the object's weight must be constant.
Student 2: That is apparently correct, but can we really assume that gravity is constant? I thought that it changed depending on your location on the planet.

Scenario 5) Two inexperienced software developers.
Developer 1: Is there a logic function that requires _______ A nor B to be true to trigger an output?
Developer 2: Yes, I think it is the Boolean function, NAND.
Developer 1: Are you sure? I thought it was NOR.

Scenario 6) A city planning office.
Traffic engineer: At that junction the drivers have a choice between taking a short distance route ________ taking a longer distance route that is actually faster. At 7.45am most drivers prioritize speed so that they can get to work on time. _______ most take the second route – leading to predictable traffic congestion.
His boss: There will be a review of the transport and traffic strategy for the city this autumn. This is just the type of data that will be useful for that process.

Scenario 7) Sports commentator currently describing a race.
The Olympic record for 800m is 1 minute 35 seconds but the World record is 1 minute 33 seconds. If Kip Scott keeps running at this pace, he will finish the first lap in 47 seconds and will have a real chance to break the world record. Neither the second nor the third place runner are even close to him _________ they have no chance of breaking __________ World __________ Olympic records.

UNIT 18: CONNECTIVES - CAUSE

Technical, scientific and medical English frequently require an explanation of events. In essence, these connectives are used to explain ***why*** something has happened. The following are common examples of connectives used in written and spoken English.

Due to (F) = Caused by. This expression is used to describe the triggering or initiating event.
Due to is usually preceded by some tense of the verb ***to be***. For example:
The delay to the train was **due to** excessive snowfall.
Rainbows are **due to** the refraction of light by raindrops.
His high temperature was **due to** a bacterial infection.
Cell growth is **due to** alterations in gene transcription that are, in turn, due to hormone-receptor binding.
What if global warming is not **due to** carbon dioxide emissions?
Due to her hard work and application, she excelled in my tutorials and eventually came top of her class.

Caused by (N) = Due to. **Caused by** cannot be used at the beginning of a sentence. For example:
Many diseases are **caused by** genetic mutations, very few are purely caused by environmental factors.
Down's syndrome is **caused by** trisomy of chromosome 21.
The road traffic accident was **caused by** a driver who had been drinking too much.
The release of an antineutrino was **caused by** the decay of a neutron.
The release of carbon dioxide was **caused by** the addition of acid to bicarbonate.

Because of (N) = The reason for, the explanation for. **Because of** is very commonly used in speech.
Because of can be used at the beginning of a sentence. For example:
Because of its mass of approximately two units, the unknown particle cannot be a proton.
Because of the age of the discovered artifacts, the archaeologist stated that the site originated in the palaeolithic era.
Are men and women attracted to each other **because of** chemicals like pheromones?
I asked for her opinion **because of** her expertise in this area.

On account of (F) = As a consequence of, As a result of. For example:
On account of the lack of success of the engineering department in obtaining funding grants, a major personnel review will be carried out.
This plane has a greater range **on account of** its greater fuel capacity.

As a result of (F) = As a consequence of. This can be used at the beginning of a sentence. For example:
As a result of its recruitment and training, Harvard produces excellent scientists.
Will the population decrease **as a result of** the smaller birth rate per capita (per person)?

> Another variation that is often used to explain events, makes use of "reason" and "cause." Specifically, **the reason for** and **the cause of** are commonly used introductory phrases.

The reason for (N) = The explanation for. This expression introduces an interpretation and explanation of events.
The cause of (N) = The trigger for, the initiator of. This expression introduces a statement explaining why events *started*. For example:
The reason for the greater prevalence of melanoma is increased exposure to the sun.
Is **the reason for** increased levels of crime, more reporting or less spending on the police?
The reason for the slow reaction is the poor mixing of the aqueous and organic phases.
The primary **cause of** gastrointestinal cancers is poor diet.

(N) = Neutral, (F) = Formal

UNIT 18: CONNECTIVES - CAUSE (EXERCISES)

Because	Due	Reason	Cause	Result	Account	Caused

A) Complete the following sentences with one of the above words:

1) The rise in body temperature was ___________ by decreased conduction, convection and radiation.

....

2) The lifetime of this solar system is limited ____________ of the nuclear reactions occurring in the sun.

....

3) On __________ of his acute and severe blood loss, the patient suffered multi-organ failure and died.

....

4) The _______ for the decrease in the number of this species, is the increase in the number of its predators.

....

5) The _________ of the patient's encephalitis was a herpes virus infection.

....

6) An object accelerates ________ of the resulting force acting on it.

....

7) Her cancer was _________ by an inherited ras oncogene.

....

B) For each of the scenarios below, complete the sentences with the appropriate connective(s):

Scenario 1) A patient talking to his General Practitioner.
GP: Why are you worried about these headaches?
Patient: I'm worried because they are the worst headaches that I have ever had. Furthermore I think that these headaches are, in fact, migraines because they seem to be ___________ by chocolate. I read in a magazine that chocolate can cause migraines.

Scenario 2) A discussion between two car engineers in a workshop.
Engineer 1: I was very happy with the design of the car but we still lost the race.
Engineer 2: Yes, it was very disappointing.
Engineer 1: Why was their car faster than ours?
Engineer 2: I imagine it was __________ of their more efficient engine and a better drag coefficient.

Scenario 3) Newspaper article – the economics section.
On ________ of the stagnant economy and poor growth figures, a change in the interest rates is likely.
However some commentators believe that the _________ for the change in interest rates is actually the government's concern about the approaching elections.

Scenario 4) Organic chemistry postgraduate students talking to each other in a canteen.
Student 1: How can we synthesize this compound?
Student 2: I suggest that we start with a biological molecule that can be altered for our purposes.
Student 1: Why?
Student 2: ___________ of its structure, fewer steps are required to complete the synthesis. So it is a cheaper synthetic route.

Scenario 5) Two computer technicians talking during their coffee break.
Technician 1: Why did you buy that computer? It seems expensive.
Technician 2: The _______ for my purchase becomes clear if you consider the following. It has a large hard disk that is a flash drive, it has 32GB of RAM and it comes with a five year guarantee. The real question is, why didn't you buy it?
Technician 1: Because I didn't run over my old computer with my car.

UNIT 19: CONNECTIVES - CONTRAST

But (N) = However. **But** indicates a contrast or exception. **But** is used for a stronger level of contrast than *however*. For example:

"The molecular weight of phenol is 94.11, **but** the molecule that you have synthesized has a molecular weight of 97.5. Perhaps you should try again."

"Your curriculum vitae indicates that you have expertise in genetics and virology, **but** you have failed to show any such skills in the last 2 years."

"According to the manufacturer, this equipment is designed to tolerate temperatures of up to 50°C. **But,** in my experience, you start to get problems at about 35 °C."

"The surgeon has removed the right leg **but** the medical notes indicated that the left leg was scheduled for removal!"

However (N) = In contrast, but. The contrast is not as marked as with *but*. **However** is also a more socially acceptable way of stating disagreement. For example:

"He did not pass his final biochemistry examination. **However**, he only failed by a small amount.

She does not have a degree in mechanical engineering; **however**, she does have a degree in applied physics."

"Organic light emitting diodes (OLEDs) may soon replace light bulbs as the former are more energy efficient. **However**, their positive green credentials are balanced by the facts that OLEDs are expensive to make and have short lives."

> Notice the different grammatical constructions used by **but** and **however**:
> The plant requires temperatures above 30 °C**. However,** the temperature here is 28 °C.
> The plant requires temperatures above 30 °C**; however,** the temperature here is 28 °C.
> The plant requires temperatures above 30 °C **but** the temperature here is 28 °C.

In contrast (to) (F) = Contrary to. **In contrast to** is used to compare situations that are in disagreement. For example:

"There are two competing hypotheses to explain this phenomenon. The first assumes that fluid mechanics is applicable. **In contrast**, the second hypothesis makes no such assumption."

"In the UK, ten people a day die from alcohol poisoning, **in contrast** one person a day die from the drug ecstasy (MDMA). So why is ecstasy illegal but alcohol legal?"

> **A more *informal* version of "**in contrast**" is "**on the other hand.**" The combination of** on the one hand **and** on the other hand **is often used in conversation.**

"**On the one hand**, bacteria are a cause of disease and death throughout the world and so we make a great effort to eliminate bacteria. **On the other hand**, we have an absolute requirement to maintain bacteria in our bodies in order to sustain life."

Contrary to (F) = In conflict with, despite. For example:

"The actual performance of this microprocessor is **contrary to** the predictions of the research and development team. It is signficantly slower than we expected."

"Metronidazole and cefuroxime are usually an effective antibiotic combination for post-surgical patients. However, recently the rates of mortality have been increasing, **contrary to** expectations."

Yet (N) = However, nevertheless. **Yet** is often used to introduce an unexpected contrast. For example:

"If your hypothesis is correct this reaction should produce three protons and a neutron. **Yet** our measurements indicate only two protons and a neutron."

"The electrical wiring in this house and all the electrical fixtures are brand new. **Yet** every two weeks one of the main fuses blows."

"We have tried calcium antagonists, beta blockers, ACE inhibitors and diuretics, **yet** his blood pressure remains intractably high."

(N) = Neutral, (F) = Formal

UNIT 19: CONNECTIVES - CONTRAST (EXERCISES)

Yet	**Contrary to**	**contrast to**	**However**	**But**	**On the other hand**

A) Complete the sentences below with the appropriate connective or connective phrase from above. For example:

His performance in the physics assessment was in ______ ______ his performance throughout the rest of the academic year.

His performance in the physics assessment was in **contrast to** his performance throughout the rest of the academic year.

1) Using V=IR we would expect the voltage drop across this component to be 6 mV; ________, it is 12 mV.

....

2) "The operation was a success ___ the patient died." Maybe you need to reconsider what you call success!

....

3) Until recently tetrahydrides were being synthesized for use as superconductors. __________, the latest research indicates that trihydrides can act as superconductors at lower more attainable pressures. This is directly in _________ _________ work done in our laboratory, which indicates that trihydrides are not good superconductors.

....

....

....

4) ______ _______ expectations the temperature of the city fell more than that of the local countryside last Monday. Why didn't the urban heat island effect apply then?

....

....

5) On the one hand, we are under pressure to deliver good and economical patient services by using computers. _____ _______ ______ _______ we are not being given the money, equipment or legislative support to ensure widespread use of computers in medicine.

....

....

6) This species was declared extinct ten years ago. ______ during our expedition in Chile we came across a freshly decomposing male of the species.

....

....

7) Currently, rapid wireless communications make use of radiowaves as the transmission medium. _______, because radiowaves can pass through buildings, the information can be intercepted and so is not secure. In _______ ___ this, the latest optical broadband would be more secure because it is restricted to inside the building. Additionally, the optical technology has a greater bandwidth.

....

....

....

8) __________ _________ expectations, long term oral ingestion of antibiotics actually predisposes an individual to secondary bacterial infections.

....

9) On the one hand patients with depression have an increased risk of mortality ______ ______ _______ _______ patients with depression and anxiety have a lower mortality compared with depression alone.

....

....

10) __________ ________ current standards, in the future all nuclear reactors will be based on fusion rather than on fission.

....

UNIT 20: CONNECTIVES - COMMENT

The following words are frequently used when giving an opinion. They are usually placed at the beginning of the relevant sentence or phrase, for example:

"***Obviously,*** if you eat too much you will become too big."

Such words are commenting on the opinion that is being offered. In this example, the speaker is indicating that she or he believes the results of overeating are clear and predictable.

Obviously (N) Clearly.

Obviously, your theory is not supported by the facts.

This program was produced very quickly and is full of errors, we **obviously** cannot use it to run the power station.

From the way that the athlete was walking he was **obviously** injured.

Cambridge University has more alumni with Noble prizes than any other university in Europe. **Obviously**, Cambridge's reputation and its competitive selection process have contributed to the success of its alumni.

Naturally (N) To be expected, a natural consequence of an early statement or event.

Naturally, I would be delighted if my paper was accepted for publication in Nature.

Naturally, the more money you invest in development, the better the final technology.

Fortunately (F) Luckily.

Fortunately, when it crashed, the rocket was not carrying any passengers.

Fortunately, for sufferers from Parkinson's disease, adult brains can accept nerve cells from human embryos.

Fortunately, the closed circuit television automatically recorded the face of the thief.

Unfortunately (F) Unluckily.

Unfortunately, she was not resistant to that form of malaria.

Unfortunately, I did not see last year's lunar eclipse.

Unfortunately, the nuclear power station released radioisotopes all over the country.

Of course (N) Obviously, naturally (often used to agree with a previous statement).

Of course, if he had been wearing safety glasses then his eyes would be fine.

Of course, all of the machinery will have to be well maintained and replaced when necessary.

Of course, further experiments will have to be performed to confirm our suspicions.

Actually (N) In reality, in fact, really.

Actually, Einstein's theory of relativity was an oversimplification.

Actually, this year 50% of the accepted applicants were women rather than the anticipated 30%.

Although you would expect many individuals to be both competent doctors and scientists, it is **actually** very rare to find an individual that excels in both fields.

Frankly (I) Honestly, plainly (usually used to express anger, irritation or impatience).

Frankly, I have no idea what you are talking about.

Quite **frankly,** the design of the engine is secret and will not be released to you or any other journalist.

(I) = *Informal*
(N) = *Neutral*
(F) = *Formal*

UNIT 20: CONNECTIVES - COMMENT (EXERCISES)

For each of the scenarios below, complete the sentences by choosing the best connective:

Scenario 1) A surgeon describing the results of his first attempt at a cardiac transplant, to his boss.

The heart we used was harvested from a young donor in a persistent vegetative state. It was brought to our hospital by helicopter. I myself examined the heart to confirm that it was in good condition. The operation itself went very smoothly and only took three hours. The recipient was cardiovascularly stable throughout the procedure. There were no complications. *(Fortunately/Unfortunately/Naturally),* although the operation was a success, the patient died. We have requested an autopsy to find out why the patient died.

Scenario 2) A postdoctoral scientist talking to her husband at the end of a busy day.

It has been a tough day. I repeated the experiment 5 times. I tried changing all the reagents and most of the equipment and I still didn't get the result I was expecting. *(Obviously/Unfortunately/Fortunately),* I will try again tomorrow but I am not hopeful.

Scenario 3) Architect talking to a project manager.

The possible renewable energy sources we considered were wind energy, solar energy, geothermal energy and wave energy. Each has its strengths and weaknesses. *(Frankly/Naturally/Actually),* in a region with abundant sunlight throughout the year, solar energy is preferred. *(Obviously/Fortunately/Frankly),* in an area with less direct sunlight and more wind, wind energy would be a better choice.

Scenario 4) Botanist lecturing a student.

Photosynthesis requires light to convert carbon dioxide and water into carbohydrate and oxygen. Although this is the accepted dogma, *(actually/obviously/ frankly)* not all the frequencies of visible light are necessary for this process. For example, green chlorophyll has a different light absorption spectrum from brown chlorophyll but they can both support photosynthesis.

Scenario 5) Human Resources Manager discussing job applications with a colleague over lunch.

We use multiple choice tests to assess the candidates and these results play a major role in the selection process. Is it appropriate for a written test to be so important in the selection process? Has this particular written test even been properly validated? *(Frankly/Naturally/Unfortunately),* I believe strongly that the candidate's past performance in real work situations should be given greater consideration than a test.

Scenario 6) School chemistry teacher talking to his/her class.

Most people believe correctly that sulphuric acid is a strong acid. However non-scientists frequently confuse corrosiveness with strength. *(Actually/Naturally/Obviously),* sulphuric acid is a strong acid because it dissociates freely in water and not because it is corrosive. *(Of course/Frankly/Unfortunately),* acid strength correlates positively with corrosiveness.

Scenario 7) Extract from a journal article.

Pure water solidifies at about 0°C under normal atmospheric pressure. Similarly, ice melts at 0°C. As the temperature is raised, the energy of the water molecules increases until they can go from the liquid state to the gaseous state. *(Fortunately/Obviously/Actually)* then, the boiling point of any liquid will higher than the melting point of the same material.

Scenario **8) A forensic pathologist giving evidence at a trial.**

Fingerprint comparison requires a twelve point match for identification. We recovered four fingerprints from the scene of the crime. *(Unfortunately/Naturally/Fortunately),* in this case we only have six point matches, so the identification is unreliable.

UNIT 21: SPECULATION

Conditional sentences are often used either for speculation or to describe natural events/phenomena/automatic events. Conditional sentences can be divided into two parts; the "**if**" clause and the "**main**" clause.

A) Type 1 conditional sentences ("Probable")

The basic structure of a type I conditional sentence:
The verb in the "**if**" clause is usually in the **present tense** and the verb in the "**main**" clause is in the **future tense**.

"If **you want** to decrease experimental error, *you* **will have to** *improve your technique*."

"You **will have to** improve your technique, if **you want** to decrease experimental error."

Reversing the position of the if and main clauses does not alter the meaning of the sentence.

B) Exceptions to Type 1 conditional sentences:

"If you **put** too much current through a resistor it **produces** heat and burns."
"If rubber **is stretched** a little, **it extends**."
"If you **lower** the temperature of the metal it **becomes** a superconductor."
"If you **add** antioxidants to the diet you **can** decrease the effects of aging."
"If the oil supply **decreases,** petrol prices **go up**."

The above are examples of type I sentences that use **two present tenses** in sequence. It is more usual for technical subjects to use this construction to describe natural or automatic events. However it is still possible to use the normal type I construction:

"If you put you too much current through a resistor it **will** produce heat and burn."
"If rubber is stretched a little, it **will** extend."
"If you lower the temperature of the metal it **will** become a superconductor."
"If you add antioxidants to the diet you **will** decrease the effects of aging."
"If the oil supply decreases, petrol prices **will** go up."

C) Type 2 conditional sentences ("Improbable")

The basic structure of a type 2 conditional sentence:
The verb in the "**if**" clause is usually in the **past tense** and the verb in the "**main**" clause is in the **conditional tense**.

"If he **became** Head of the Department, I **would** resign.'

Again, reversing the if and main clauses does not change the meaning. Generally, the scenarios being described in the type 2 conditionals are less likely than in the type 1 conditionals. *The statement in the if clause is usually unlikely.* So in the above example, the speaker is assuming that it is unlikely that ***"he"*** will become head of department.

"If 100W light bulbs **were used** in this building the basic electricity costs **would increase** by 20%."
"If we **launched** the rocket after 6pm it **would** arrive too late to dock with the satellite."
"If the tram network **extended** to this hospital, the staff **would** avoid the traffic and arrive at work on time."
"If they **carried** out that orthopaedic surgery on a bacteraemic patient, there would be a significant risk of osteomyelitis and septic arthritis."
"If you routinely **treated** childhood cancers with radiotherapy, the incidence of adult cancer **would** rise."
"What would you do if the budget for the department increased by 30%?"
"If the budget for the department increased by 30% we would employ more personnel."

Any conditional sentence can be rephrased as a question:
If the tram network extended to this hospital, would the staff avoid traffic and arrive at work on time?
If you routinely treated childhood cancers with radiotherapy, would the incidence of adult cancer increase?
If you lower the temperature of the metal, does it become a superconductor?
If you lower the temperature of the metal, what does it become?

UNIT 21: SPECULATION (EXERCISES)

A) Rephrase the conditional statements below as questions:

For example: If the rocket keeps moving at this velocity it will arrive at the moon in 6 hours.
If the rocket keeps moving at this velocity will it arrive at the moon in 6 hours?

1) If you drive to Paris your carbon footprint will be less than taking a plane.

....

2) If your iodine intake is too high you will be prone to hyperthyroidism.

....

3) If you look directly at the sun through a telescope, you will be blinded.

....

4) If everybody on the planet were a vegetarian there would be enough food for everyone.

....

5) If sales of the New Scientist increased at that rate they would overtake the Scientific American.

....

6) If we lost the ozone layer, life on the planet would cease.

....

B) Complete the sentences below with the suggested verb in the appropriate tense:

For example: If you increase the dose of digoxin patients (die).
If you increase the dose of digoxin patients will die.

1) If we (dig) beneath the layer of sedimentary rock, what do you think we will find?

....

2) If she were exposed to Lancefield Group A beta haemolytic streptococci in her current state, what (happen)?

....

3) If we alter this part of the molecule, it (be) effective at the receptor?

....

4) Would the pipes burst if we (increase) the gas pressure?

....

6) What (happen) if the government were to cut public spending by 20%?

....

7) If you (increase) the carbon content of this steel would it be stronger?

....

8) If you halved the number of DNA repair enzymes, there (be) more cancers?

....

9) The digital revolution has greatly increased the use of cameras. If there (be) an optical revolution will there be a similar increase in video use?

....

C) Using these sentences as examples, change the following sentences into speculative questions:

For example: What would happen if (raising temperature of the sun by 500°C).
What would happen if **we raised** the temperature of the sun by 500°C?
What will happen if (decreasing traffic flow on this road).
What will happen if **we decrease** traffic flow on this road?

1) What would happen if (making a plastic plane).

....

2) What will happen if (making a drug for autistic people that increases emotional sensitivity).

....

3) What will happen if (making silicon chips dosed with germanium).

....

4) What would happen if (making more efficient DNA repair enzymes).

....

UNIT 22: NUMBERS

A) Cardinals

These numbers are used to describe absolute countable quantities:

Number	Words	Number	Words
1	One	**22**	Twenty-two
2	Two	**23**	Twenty-three
3	Three	**24**	Twenty-four
4	Four	**30**	Thirty
5	Five	**40**	Forty
6	Six	**50**	Fifty
7	Seven	**60**	Sixty
8	Eight	**70**	Seventy
9	Nine	**80**	Eighty
10	Ten	**90**	Ninety
11	Eleven	**100**	Hundred
12	Twelve	**201**	Two hundred and one
13	Thirteen	**300**	Three hundred
14	Fourteen	**1000**	Thousand
15	Fifteen	**2000**	Two thousand
16	Sixteen	**3000**	Three thousand
17	Seventeen	**10,000**	Ten thousand
18	Eighteen	**100,000**	Hundred thousand
19	Nineteen	**1,000,000**	Million
20	Twenty	**1,000,000,000**	Billion (US)
21	Twenty-one	**1,000,000,000,000**	Billion (UK)

For example: "There are only **four** nucleotide bases in DNA but several **million** possible coding sequences."
"NASA spends more than **three billion** dollars on the space program each year."
"The gross domestic product of this country is **40 billion** dollars per year."

B) Ordinals

These adjectives are used to describe the order of a sequence:

Abbreviated ordinal	Full form	Abbreviated ordinal	Full form	Abbreviated ordinal	Full form
1st	**First**	17th	**Seventeenth**	80th	**Eightieth**
2nd	**Second**	18th	**Eighteenth**	90th	**Ninetieth**
3rd	**Third**	19th	**Nineteenth**	100th	**Hundredth**
4th	**Fourth**	20th	**Twentieth**	101st	**Hundred and first**
5th	**Fifth**	21st	**Twenty-first**		
6th	**Sixth**	22nd	**Twenty-second**	102nd	**Hundred and second**
7th	**Seventh**	23rd	**Twenty-third**		
8th	**Eighth**	24th	**Twenty-fourth**	103rd	**Hundred and third**
9th	**Ninth**	25th	**Twenty-fifth**		
10th	**Tenth**	26th	**Twenty-sixth**	1000th	**Thousandth**
11th	**Eleventh**	27th	**Twenty-seventh**	1,000,000th	**Millionth**
12th	**Twelfth**	30th	**Thirtieth**	1,000,000,000th	**Billionth**
13th	**Thirteenth**	40th	**Fortieth**		
14th	**Fourteenth**	50th	**Fiftieth**		
15th	**Fifteenth**	60th	**Sixtieth**		
16th	**Sixteenth**	70th	**Seventieth**		

For example: "This is the **fifth** time that I have applied for a research grant from this organization."
"The **sixtieth** patient to enter the clinical trial died."
"The atomic clock can measure a second to the **seventh** decimal point."

UNIT 22: NUMBERS (EXERCISES)

A) Are the following statements true or false?

For example: 1st = first. TRUE

100,000 is a billion. FALSE

1) 16th = sixth ....
2) 1000 is a thousand ....
3) 105 is a hundred five ....
4) 70th = seventieth ....
5) 26th = twenty-six ....
6) 2002 is twenty and two ....
7) 37 = thirty-seventh ....
8) 1000th = thousandth ....
9) 400 is four hundred ....
10) 104th = one hundred and forty ....
11) 27th = twenty-seventh ....
12) $1{,}000{,}000^{th}$ = millionth ....
13) 1001 is one thousand one ....

B) Express the following numerical expressions as words.

For example: The number of Clinical Excellence Award holders in this region is **60**.
The number of Clinical Excellence Award holders is **sixty**.

1) This is the **24th** month of economic recession.

....

2) There are **800** dentists in Portugal.

....

3) **12,000,000** women aged between 18 and 64 are screened for cervical cancer in the UK.

....

4) Is Mars the **7th** planet from the Sun? No.

....

5) If you try to open this combination lock, you would be unlikely to be successful before your **10,000th** attempt.

....

6) What is the **12th** number in this geometric progression?

....

7) A mile is **1600** metres.

....

8) This high speed train has the capacity to carry **2000** passengers.

....

C) Write the following words as numerical expressions.

For example: There are more than a **billion** grains of sand on this beach.
There are more than **1,000,000,000** grains of sand on this beach.

1) The Earth is about **ninety-three million** miles from the sun.

....

2) He shuffled the pack of cards and the **tenth** card that he drew was an ace.

....

3) The probability of the **fiftieth** coin toss producing another head is 1/2.

....

4) Every year the country produces **three thousand** new doctors and **five hundred** new consultants.

....

5) If the car accelerates from zero to a **hundred and fifty** miles per hour in six seconds, what is the average acceleration?

....

6) The pressure can be calculated as a **hundred** newtons divided by the area of ten square centimetres.

....

7) Light travels at approximately **three hundred million** metres per second.

....

8) This is the **twenty-second** population census that our company has carried out.

....

UNIT 23: NUMBERS II

A) Reading and writing fractions:

Irregular fractions:		Regular fractions:
1/2 a *half*		**1/6 a sixth**
1/3 a *third*	**2/3 two thirds**	**1/7 a seventh**
1/4 a *quarter*	**3/4 three quarters**	**1/11 an eleventh**

1/5 a fifth	1/11 an eleventh	1/17 a seventeenth	*1/100 a hundredth*
1/6 a sixth	1/12 a twelfth	1/18 an eighteenth	*1/1000 a thousandth*
1/7 a seventh	1/13 a thirteenth	1/19 a nineteenth	*1/1000000 a millionth*
1/8 an eighth	1/14 a fourteenth	1/20 a twentieth	*1/1000000000 a billionth*
1/9 a ninth	1/15 a fifteenth		
1/10 a tenth	1/16 a sixteenth		

It can be seen that, generally, any fraction of a whole number n, such as 1/**n**, is read as a(n) **n**th

Plural fractions

Consider the following examples:

3/7 three sevenths
5/11 five elevenths
6/13 six thirteenths

1/5 a fifth => 2/5 two fifths => 3/5 three fifths => 4/5 four fifths

i) It is possible to use **one** in place of **a(n)**. For example, one twelfth is the same as a twelfth.
ii) Although it is possible to say *two quarters* or *three sixths* and other similar fractions, it is more usual to say *a half.*
iii) Note that we say a fi**fth** and not a fi**veth**, similarly we say a ni**nth** not a nin**eth**, we say a twel**fth** not a twel**veth** and we say an eigh**th** and not an eigh**tth**.

B) Standard form

Fractions written in standard form can be read as follows:

1×10^{-9} One **times ten to the (power of) minus** nine
1×10^{-6} One **times ten to the (power of) minus** six
1×10^{-3} One **times ten to the (power of) minus** three

Orders of magnitude can also be read in the same manner:

10^{8} Ten to the (power of) eight
10^{-4} Ten to the (power of) minus four

C) Reading and writing large numbers

Consider these examples:

12,001,006 => Twelve million, one thousand and six
6,100,040 => six million, one hundred thousand and forty
576,123,456 => Five hundred and seventy-six million, one hundred and twenty-three thousand, four hundred and fifty-six.
8,000,900,400 => Eight billion, nine hundred thousand, four hundred.

These are examples of words being converted into numbers:

"I just wrote a cheque for nine million, six hundred thousand dollars!" => **$9, 600,000**

"I phoned you to tell you that the expenses bill you submitted for the New York conference was too large; it was seven thousand, six hundred and thirty-two pounds." => **£7,632**

"This institute's research spending is seven million, five hundred thousand pounds per year." => **£7,500,000**

"The accountants have calculated this hospital's spending on antibiotics. It is three hundred and forty-seven thousand pounds this year." => **£347,000**

UNIT 23: NUMBERS II (EXERCISES)

A) Are the following statements true or false? For example:

1/4 is a quarter. TRUE
1/5 is a **five**th. FALSE

1) 2/3 is two thirds	....	8) 3/7 is three sevenths	....
2) 1/13 is a twelfth	....	9) 2/3 is two thirds	....
3) 1/6 is a fiveth	....	10) 1/2 is a halve	....
4) 3/4 is three quarters	....	11) 1/2 is a secondth	....
5) 1/11 is an eleventh	....	12) 1/20 is a twentieth	....
6) 1/5 is a fifth	....	13) 1/13 is a thirteenth	....
7) 1/4 is a fourth	....	14) 1/8 is an eighth	....

B) How would you express the following fractions and numbers as words? For example:

12 3/4 is twelve and three quarters. 5 1/2 is five and a half.

1) 1 1/2	....	7) 2 1/5	....
2) 3 5/6	....	8) 3 2/3	....
3) 7 2/3	....	9) 3 1/2	....
4) 4 7/8	....	10) 12 4/5	....
5) 2 3/11	....	11) 11 6/7	....
6) 1 3/7	....	12) 10 5/8	....

C) Change the following words into numbers and fractions. For example:

A quarter is a 1/4. Three sevenths is 3/7.

1) Three quarters is	....	7) Four and a fifth is	....
2) A half is	....	8) Three sevenths is	....
3) Seven eighths is	....	9) Eight and five ninths is	....
4) Two thirds is	....	10) Twenty and three quarters	....
5) Twelve and a quarter is	....	11) A third is	....
6) Five sixths is	....	12) Seven and two sevenths is	....

D) Write the following expressions as numbers *or* expressions as words. For example:

Ten to the power of eight is $\mathbf{10^8}$.
1 x 10^{-9} is **one times ten to the minus nine**.

1) The speed of light in a vacuum is $\mathbf{3 \times 10^8}$ metres per second.

....

2) The length of a human life is about **two times ten to the nine** seconds.

....

3) The radius of a proton is about $\mathbf{10^{-15}}$ metres.

....

4) The diameter of a small bacterium is about **ten to the minus six** metres.

....

5) Is the mass of the Earth about $\mathbf{10^{24}}$ kg?

....

6) I think the Avogadro constant is $\mathbf{1.66 \times 10^{23}}$ mol^{-1}.

....

7) A resistance of 5 kΩ is **five times ten to the three ohms.**

....

8) A solution of pH 3 contains **ten to the minus three** molar H^+.

....

9) There are the meanings for about $\mathbf{2 \times 10^5}$ words in the dictionary.

....

10) The mass of the Sun is about **ten to the power of thirty** kilograms.

....

UNIT 24: ELEMENTARY CALCULATIONS

A) Basic mathematical symbols

Each of the symbols has more than one word that can be used to describe it:

Addition (+)	Subtraction (-)	Multiplication (x)	Division (÷)
Plus	Minus	Multiplied by	Divided by
And	Subtract	Times	Into (**I**)
Add (I)	Subtracted from		
Add on (I)	Take away (**I**)		
	Less (**I**)		

(**I**) = Informal

For example:

4 + 7 Four plus seven

3 x 4 Three times four

7 ÷ 2 Seven divided by two

7 ÷ 2 Two into seven

12 - 6 Twelve minus six
or Twelve subtract six
or Six subtracted from twelve
or Twelve less six
or Twelve take away six

B) Further mathematical symbols

The alternative words used to describe each of the symbols, is given below. (**I**) refers to an **i**nformal context.

Equality (=)	**Inequality (<)**	**Inequality (>)**
Equals	Less than	More than
Equal	Smaller than	Greater than
Are (**I**)	**Inequality (≤)**	**Inequality (≥)**
Is (**I**)	Less than or equal to	More than or equal to
Make (**I**)	Smaller than or equal to	Greater than or equal to
Makes (**I**)		
Goes* (**I**)		

* Used only in division

For example:

$3 + 7 = 10$ Three plus seven equals 10

$B \times D > F$ B times D is greater than F

$E - 9 \geq 3$ E minus nine **is** greater than or equal to 3

$6 \div 2 = 3$ Two into six goes three

$8x = 65$ Eight x equals sixty-five

C) Other important symbols:

∞ Infinity

ø Null set

≡ Defined as

: to *(for ratios)*

≈ Approximately equal to

± Plus or minus

|A| The modulus of A or the absolute value of A

For example:

13 ± 3 Thirteen plus or minus three.

$A.B \equiv F$ A B is defined as F

$|A|$ The modulus of A

4:1 *(A ratio of)* four to one

UNIT 24: ELEMENTARY CALCULATIONS (EXERCISES)

A) Write out the following mathematical expressions in words. For example:

78 x 5 **Seventy-eight multiplied by five** (or seventy-eight times five)
12 ÷ 4 **Twelve divided by four** (or informally; four goes into twelve....)

1) 567 + 98	....	7) 3600 x 0.5	....
2) 89 - 75	....	8) 110 - 6.219	....
3) 34 x 2	....	9) 1200 ÷ 0.7	....
4) 50 ÷ 111	....	10) 234 + 4 > 89.780	....
5) 6 x5 > 4 x 3	....	11) 0.285 - 45 + 9.34	....
6) 965 - 43	....	12) 0.78 + 4.25 + 6000 - 4.5	....

B) Change the following words into simple mathematical expressions.

For example: One thousand times fifty. **1000 x 50**
Sixteen point four minus seven. **16.4 – 7**

1) Eighty-nine add on five..
2) Fifty-six subtract eighteen...
3) Twelve times thirty..
4) Five into fifty goes ten...
5) Forty-five divided by three...
6) Two hundred plus thirty...
7) Six thousand minus one thousand two hundred...
8) One million and one plus one..
9) Twelve plus six, divide by three and subtract six...
10) Five point six plus two point seven two one eight...
11) Seventy-two thousand times four...
12) One million five hundred thousand take away two hundred.................................

C) Examine the following examples:

Add about fifty more sheets to the total........................ **+50**
We get a discount of 10% for buying in bulk, that should take 200 pounds from the price..... **-200**

Do the same for the following sentences:

1) The delay adds twenty minutes to your arrival time...
2) If you remove the VAT (value added tax) then the car costs four hundred and twenty pounds less.......
3) This computer would need two more gigabytes of memory to run this program............................
4) If we divide the work between the six of us the project should take three months...........................
5) It will take a graduate student three times as long as a senior scientist to do the research................
6) The average cat weighs five kilograms, plus or minus one kilogram...

D) Are the following statements the same? For example:

20.1 ± 5 <=> Twenty point one plus or minus five. ...True
19.1 ≈ 20 <=> Nineteen point one is approximately equal to twenty-**two** ...False

1) F ≡ MA <=> F is defined as M times A.
2) 20 x 20 = 400 <=> Twenty times twenty makes four hundred.
3) Six multiplied by five is less than forty. <=> 6 x 5 > 40
4) Fifteen plus seven divided by eighteen is less than ten. <=> (15 + 7) ÷ 18 < 10
5) G = 6.7 x 10^{-11} <=> G is equal to six point seven times ten to the minus eleven.
6) 0.8 = 80% <=> Zero point eight equals eighty per cent.
7) Fifty plus or minus six. <=> 50 ± 6

UNIT 25: RATIOS, PERCENTAGES AND DECIMALS

A) Simple Ratios

The ratio 1:10 can be spoken in different ways:

1 in 10 ...."Here, **one in ten** individuals are carriers of the sickle cell gene."
"A hill with an inclination of **one in ten** does not require a bicycle lift."
1 out of 10......................"**One out of ten** applicants are successful."
A ratio of 1 to 10..........."The reactants must be mixed in a molar ratio of **one to ten**."
1 to 10..........................."The pastry requires a **one to ten** ratio of water to flour."

B) Percentages

Percentages are said as the appropriate **number + per cent (of)**, for example:

"**Twenty per cent** of the national income is spent on the military."
"Asthmatics have lost more than **ten per cent of** their respiratory peak flow."
"What percentage of the "genome" is in fact protein? I'm not sure but it could be as much as **thirty per cent of** the genome."

The word **percentage** is used generally and is not used directly next to a number. For example – "What **percentage** of graduate students publish five or more publications?" "What **percentage** of trains are still in service after 20 years? **Fifteen per cent**."

For more complex percentages:

10.654% ⇨ Ten point six five four per cent
0.579% ⇨ Zero point five seven nine per cent ***or***
Point five seven nine per cent (the zero can be omitted if it occurs alone before the decimal point).
129.23% ⇨ One hundred and twenty nine point two three per cent ***or***
One two nine point two three per cent (this is a less common construction which may, for example, be used over the telephone).
78.05% ⇨ Seventy-eight point zero five per cent ***or***
Seventy-eight point nought five per cent
25.0% ⇨ Twenty-five point zero per cent ***or***
Twenty-five point nought per cent ***or***
Twenty-five per cent *exactly*

C) Decimals

In order to read out percentages, it is important to be able to read out the respective decimal:

37.5 ⇨ **37.5%**
Thirty-seven point five ⇨ Thirty-seven point five per cent
12.06 ⇨ **12.06%**
Twelve point zero six ⇨ Twelve point zero six per cent ***or*** Twelve point nought six per cent
0.04 ⇨ **0.04%**
zero point zero four ⇨ Zero point zero four per cent ***or*** Nought point nought four per cent
82.001 ⇨ **82.001%**
Eighty-two point zero zero one per cent ***or*** Eighty-two point nought nought one per cent

"What chance do you have of winning the Fulbright scholarship?"
"Well, because there are only three applicants, my chances are roughly **thirty-three point three per cent**."
"What will she do if she gets the promotion to Director of the Institute?"
"I think that there is a **ninety-nine point nine** per cent chance that she will stay here until she retires."

UNIT 25: RATIOS, PERCENTAGES AND DECIMALS (EXERCISES)

A) Are the following statements true or false? For example:

12% is twelve percentage. FALSE
0.34% is point three four per cent. TRUE

1) 2:5 is two by five ....
2) 5% is five per cent ....
3) 0.64 is zero point six four per cent ....
4) 0.087 is nought point nought eight seven ....
5) 1.92% is one point nine two per cent ....
6) 12.03 is twelve point zero third ....
7) 4:1 is four to one ....
8) 30.2% is thirty point two per cent ....
9) 3:4 is three out of four ....
10) 70.7% is seventy point seventh per cent ....
11) 15:2 is five to two ....
12) 33.5% is thirty-three point five ....
13) 7:1 is seven to one ....
14) 0.23 is point two three ....

B) How would you express the following numbers as words? For example:

50% is fifty per cent.
1:5 is a ratio of one in five.

1) 7.342 ....
2) 32% ....
3) 0.319 ....
4) 11.1% ....
5) 6:1 ....
6) 54.7 ....
7) 1:5 ....
8) 45.3% ....
9) 0.9356 ....
10) 30.0% ....
11) 7:2 ....
12) 89.5% ....

C) Change the following words into numbers. For example:

Twelve point one per cent is 12.1%. A ratio of two to three is 2:3.

1) Thirty-eight point five per cent ....
2) Point zero eight two ....
3) Three in five ....
4) Ninety-nine point nine nine per cent ....
5) Fifty seven ....
6) Seven to six ....
7) Twelve point zero three per cent ....
8) Three per cent exactly ....
9) One out of four ....
10) Eighty-three point two seven nine ....
11) Nought point two five per cent ....
12) Five to three ....

UNIT 26: UNITS AND DIMENSIONS

A) The following prefixes are used with units to indicate decimal multiples or fractions:

10^{12} Tera- T	10^{2} Hecto- h	10^{-3} Milli- m	10^{-15} Femto- f
10^{9} Giga- G	10 Deca- da	10^{-6} Micro- μ	10^{-18} Atto- a
10^{6} Mega- M	10^{-1} Deci- d	10^{-9} Nano- n	
10^{3} Kilo- k	10^{-2} Centi- c	10^{-12} Pico- p	

For example:

A mile is equal to 1.609 **kilo**metres. (km)
Most radio waves have a wavelength greater than **milli**metres. (mm)
Make a **nano**molar solution of insulin. (nM)

B) Basic units and their abbreviations:

The student should be aware of the following **imperial** units as they are still used by many people. Officially, Britain has adopted the metric system.

Weight	**Length**	**Area**	**Volume** (liquid)	**Volume**
Grain **g**	Inch **in**	Square inch $\mathbf{in^2}$	Pint **pt**	Cubic inch $\mathbf{in^3}$
Dram **dr**	Foot **ft**	Square foot $\mathbf{ft^2}$	Quart **qt**	Cubic foot $\mathbf{ft^3}$
Ounce **oz**	Yard **yd**	Square yard $\mathbf{yd^2}$	Gallon **ga**	Cubic yard $\mathbf{yd^3}$
Pound **lb**	Mile **mi**	Acre **a**	Peck **pk**	
Stone **st**		Square mile $\mathbf{mi^2}$	Bushel **bu**	
Quarter **qr**			Quarter **qr**	
Hundredweight **cwt**				
Ton **tn** (short ton **st**) (long ton **lt**)				

Note the following plurals:

foot (singular)	feet (plural)
stone (singular)	stone (plural)
inch (singular)	inches (plural)

C) Metric units

Weight	**Length**	**Area**	**Volume**
milligram **mg**	millimetre **mm**	square centimetre $\mathbf{cm^2}$	millitre **ml**
centigram **cg**	centimetre **cm**	square metre $\mathbf{m^2}$	centilitre **cl**
decigram **dg**	decimetre **dm**	are **are**	decilitre **dl**
gram **g**	metre **m**	hectare **ha**	litre **l**
decagram **dag**	decametre **dam**	square kilometre $\mathbf{km^2}$	decalitre **dal**
hectogram **hg**	hectometre **hm**		hectolitre **hl**
kilogram **kg**	kilometre **km**		kilolitre **kl**
tonne **tonne**			

In American English, met**re** becomes met**er**.

The plurals are formed by adding 's' to the end of the quantity, for example metre becomes metres.

UNIT 26: UNITS AND DIMENSIONS (EXERCISES)

A) Change the following units into words. For example:

Even **ng** quantities of the toxin are lethal............. **nanogram**
An inch is about 2.5 **cm**.. **centimetres**
I made up a 20 **mM** solution of sodium chloride... **millimolar**

1) The molecule is 5 **nm** long. ....
2) The satellite will be at an altitude (height) of 1336 **km**. ....
3) A radiowave frequency of 5 **GHz** was used to measure the distance from earth. ....
4) The reference book gives the average weight of a man as 70 **kg**. ....
5) Normal human muscle cells contain 30 - 40 **mM** of the amino acid, creatine. ....
6) The new drug is so potent (powerful) that it is therapeutic at about 9 **pM**. ....

B) Change the following words into units. For example:

The bacterium was 10 **micrometres** long.......................................**µm**
The fastest car moved at 0.283 **kilometres** per second...................**km**

1) Can the human tongue taste molecules at **picomolar** concentrations? ....
2) The volume of the test tube was 10 **millilitres**. ....
3) Dissolve 250 **micrograms** of the substance in 2 litres of ethanol. ....
4) The average cat weighs 5 **kilograms**. ....
5) The size of some of the water droplets is 30 **picolitres**. ....
6) The distance between London and Aarhus is 890 **kilometres**. ....

Inch	**Gallon**	**Pint**	**Grain**	**Stone**	**Ounce**	**Mile**	**Dram**
Cubic feet	**Yard**	**Cubic inch**	**Ton**	**Acre**	**Foot**	**Quarter**	

C) Match the following units to the words in the table above. For example:

oz is an **ounce**.
a is an **acre**.

1) ga is ....
2) ft^3 is ....
3) oz is ....
4) qr is ....
5) dr is ....
6) yd is ....
7) mi is ....
8) a is ....
9) in^3 is ....
10) in is ....
11) pt is ....
12) g is ....
13) tn is ....
14) st is ....
15) ft is ....

D) Are the following statements true or false? For example:

Ton is written as ft.........**False**
mm is a millimetre.........**True**

1) mg is a microgram. ....
2) lb is a pound. ....
3) ml is a millilitre. ....
4) st is a stone. ....
5) km is a kilometre. ....
6) square mile is written mi^2. ....
7) hectare is written ha. ....
8) q is a quarter. ....
9) millimolar is written mM. ....
10) qr is a quarter. ....
11) l is a litre. ....
12) cm is a centimolar ....
13) a is an acre. ....
14) gram is a gr. ....
15) cubic inch is written in^3 ....
16) mi is a mile. ....
17) kilogram is written kg. ....
18) ounce is written ou. ....

UNIT 27: UNITS AND DIMENSIONS II

In the expression "a^b" b is the **index** of a.
The plural of "*index*" is "*indices*".

A) Reading units with negative indices:

The following conventions are used when reading units. In the left hand box is the general rule and in the right hand box are some examples.

a/b **is read as a(s) per b**	**e.g.** "m/s" **is read as** "metres per second"
ab^{-1} **is read as a(s) per b**	**e.g.** "ms^{-1}" **is read as** "metres per second"
ab^{-2} **is read as a(s) per b squared**	**e.g.** "Nm^{-2}" **is read as** "newtons per metres squared"
ab^{-3} **is read as a(s) per b cubed**	**e.g.**"g/m^{-3}" **is read as** "grams per metres cubed"
ab^{-4} **is read as a(s) per b to the (power of) four**	**e.g.** "ms^{-4}" **is read as** "metres per second to the power of four"
ab^{-5} **is read as a(s) per b to the (power of) five**	**e.g.** "Cm^{-5}" **is read as** "coulombs per metre to the power of five"
etc.	

There are some exceptions:
i) When the index of the unit includes a 2 or 3 and refers to a dimension of length then the words "square" or "cubic" may be used, for example:

$moldm^{-3}$ can be read as *moles per cubic decimetre*
cm^{-3} can be read as *cubic centimetre*
Nm^{-2} can be read as *newtons per square metre.*
lb/in^2 can be read as *pounds per square inch.*

ii) Compound units are units that consist of two or more dimensions. These frequently have names given to them by convention, for example:

Js^{-1} is often read as **watt**
Cm is often read as **debye**

iii) *Informally* the "power of" may be omitted when reading units.

B) Reading units with positive indices:

Here there is no use of "per":

ab is read as a b(s)	**e.g. "Js" is read as "Joule seconds"**
ab^2 is read as a b squared	**e.g."Nm^2" is read as "Newton metres squared"**
ab^3 is read as a b cubed	**e.g."Nm^3" is read as "Newton metres cubed"**
ab^4 is read as a b to the (power of) four	**e.g."Nm^4" is read as "Newton metres to the (power of) four"**
ab^5 is read as a b to the (power of) five etc.	**e.g."Nm^5" is read as "Newton metres to the (power of) five"**
	etc.

Further Examples:
eV is read as "electron volt"
Ns is read as "newton second"

Again, under informal conditions, the "power of" may be omitted when reading units.

UNIT 27: UNITS AND DIMENSIONS II (EXERCISES)

A) Change the following expressions into words. For example:

J/m^2	=>	joules per square metre
ms^{-2}	=>	metres per second squared
$JK^{-1}mol^{-1}$	=>	joules per kelvin per mole

1) kJs^{-1}	....	7) $molm^{-3}$	....
2) Ωm^{-1} (Ω=ohm)	....	8) $Nm^{-2}kg^{-2}$	....
3) $moldm^{-3}$	....	9) kgm^{-3}	....
4) lm^{-1}	....	10) $kJmol^{-1}$	....
5) mmh^{-2} (h=hour)	....	11) Wm^{-2}	....
6) Ckg^{-1} (C=coulomb)	....	12) Cm^{-1}	....

B) Change the following words into units. For example:

moles per litre..................................	mol/l or $moll^{-1}$
kilojoules per square metre................	kJ/m^2 or kJm^{-2}
centimetres per second squared.........	cms^{-2} or cm/s^2

1) kilowatts per metre	....	7) coulombs per metres squared	....
2) micromoles per litre	....	8) joules per kilogram	....
3) centimetres per hour squared	....	9) moles per cubic metre	....
4) ohms per centimetre	....	10) megajoules per second	....
5) Joules per mole	....	11) pounds per square inch	....
6) grams per cubic centimetre	....	12) coulombs per metre to the power four	....

C) Are the following units correct? For example:

Nobody can run faster than **14 ms^{-1}** ......Correct

The density of the metal is **5.7 gcm** ......Incorrect (g/cm or gcm^{-1})

The kinetic energy of the electrons was equal to **30 electron volts**.....Correct

1) The greenhouse effect may cause a temperature increase of 1^oK/year. ....
2) Before the helicopter can fly the rotors must rotate at about 2,000 revolutions minute. ....
3) Is the pressure greater than 20 lbs per square inch? ....
4) It took 30 man hours to set up the equipment. ....
5) The fastest bird flies at 47 metres second. ....
6) The density of the solution was 1.1 grams per m. ....
7) As you climb up the mountain the temperature drops by 1^oK per 500 m. ....
8) The plutonium produces 600 ß particles per m^3. ....
9) There were 200 lions per 50 square miles. ....
10) 3.7 x 10^{10} disintegrations second is a curie. ....

Joule seconds	**Newtons per square metre**	**Electron volts**
Metres per second squared	**Counts per second**	

D) Match the following units to the words in the table above. For example:

m/s......metres per second

1) ms^{-2} ....
2) eV ....
3) Nm^{-2} ....
4) Js ....
5) ct/s ....

UNIT 28: UNITS AND DIMENSIONS III

*Please read **Units and Dimensions I** and **II** before starting this unit.*

A) Note the following table:

Physical Quantity	**Name**	**Abbreviation**
Length	Metre	m
Mass	Kilogram (Informally; "kilo")	kg
Time	Second (Informally; "sec")	s
Electric current	Ampere (Informally; "amp")	A
Temperature	Kelvin	K
Quantity of molecules/ions	Mole	mol or moles
Luminous intensity	Candela	cd
Plane angle (2 Dimensional-2D)	Radian	rad
Solid angle (3 Dimensional-3D)	Steradian	sr

B) In the SI system, the following derived units have special names:

Physical Quantity	Name	Abbreviation
Frequency	Hertz	Hz
Energy	Joule	J
Force	Newton	N
Power	Watt	W
Pressure	Pascal	Pa
Electric charge	Coulomb	C
Electromotive force	Volt	V
Electric resistance	Ohm	Ω
Electric conductance	Siemens	S
Electric capacitance	Farad	F
Magnetic flux	Weber	Wb
Inductance	Henry	H
Magnetic flux density	Tesla	T
Luminous flux	Lumen	lm
Illumination	Lux	lx

C) Commonly abbreviated units include:

kph kilometres per hour
cps cycles per second
mph miles per hour
fps feet per second
cps characters per second
psi pounds per square inch

UNIT 28: UNITS AND DIMENSIONS III (EXERCISES)

A) Change the following expressions into words. For example:

cd/mol..........	candelas per mole
A/m^2............	amperes per square metre
$rads^{-2}$............	radians per second squared
$kJK^{-1}mol^{-1}$.....	kilojoules per kelvin per mole

1) Ωs^{-1}	....	7) lxm^{-3}	....
2) kJm^{-1}	....	8) $Nm^{-2}kg^{-2}$	....
3) $moldm^{-3}$	....	9) $Wbmm^{-3}$	....
4) Vm^{-1}	....	10) kJm^{-2}	....
5) Tm^{-2}	....	11) $mWcm^{-2}$	....
6) μFkg^{-1}	....	12) Cm^{-1}	....

B) Change the following words into units. For example:

farads per litre.................... F/l or Fl^{-1}
kilojoules per square metre......kJ/m^2 or kJm^{-2}
teslas per second................. T/s or Ts^{-1}
coulomb metres.................. Cm

1) Farads per metre squared.	....	7) Milliamps per centimetre	....
2) Megajoule seconds.	....	8) Nanomoles per microlitre.	....
3) Microteslas per metre.	....	9) Lux per second.	....
4) Milliohm nanometres.	....	10) Centicandela metres.	....
5) Steradians per second.	....	11) Pounds per square inch.	....
6) Joules per metre.	....	12) Miles per hour.	....

C) Is the use of the following units correct? For example:

No insect can run faster than **30 mms^{-1}**..Correct
The density of the metal is **5.7 gcm** ....Incorrect (g/cm^3 or gcm^{-3})
The kinetic energy of the electrons was equal to **30 electron volts**.........Correct

1) Each person is allowed to take luggage of up to **50 kilos** in weight. ....
2) The magnetic flux density was **78 teslas**. ....
3) Carbohydrate has an energy content of **17.2 kilojoules gram**. ....
4) The speed limit is **90 kilos per hour**. ....
5) The concentration of creatine in the cell is **40 mmoles per dm^3** ....
6) Atmospheric pressure here is **101 kilonewtons**. ....
7) Is the temperature greater than **9 lx**? ....
8) The resistance of the wire was **5 Ω**. ....
9) Atmospheric pressure is more than **50 pascals**. ....
10) 240 volts are delivered at a frequency of **50 cp** to each house. ....
11) The plane rose at a rate of **200 feet per second**. ....
12) The car was not designed to go faster than **139 mph.** ....

UNIT 29: TIME

There are three commonly used conventions for describing the time in English:

A) The long form

The **hour** is said or written as shown:

1.00 => "***One o'clock***"
2.00 => "***Two o'clock***"
3.00 => "***Three o'clock***" etc.

Fifteen **minutes** after the hour is said or written as =>"***Quarter past***...."
Fifteen minutes before the hour is said or written as =>"***Quarter to***...."

For example: 2.15 => "***Quarter past two***"
1.45 => "***Quarter to two***"

Thirty minutes before or after the hour is said or written as =>"***Half past***...."
For example, 2.30 => "***Half past two***"

All other times before or after the hour use "to" or "past," respectively:
For example: 3.17 => Seventeen (minutes) **past** three
4.50 => Ten (minutes) **to** five

The time of day (**morning**, **afternoon** or **evening**) is said or written after the hour.
I start work at nine o'clock **in the morning**.
I finish work at half past five **in the evening**.

B) The short form

This simpler form tells the time by directly saying or writing the numbers:
For example: 4.15 **Four fifteen,** 7.37 **Seven thirty-seven**, 8.52 **Eight fifty-two**

The **morning** is indicated by **a.m.** (ante meridiem)
The **afternoon/evening** is indicated by **p.m.** (post meridiem)

For example: 1.22 **a.m.** One twenty-two **a.m.**
3.43 **p.m.** Three forty-three **p.m.**

Informally, especially in conversation, the hour alone may be given:
"I'll meet you in the restaurant at 7."
"Let's continue the discussion at 3."

C) The 24 hour clock

Formally, the police, lawyers and military frequently use the **24 hour clock**:

16.00	sixteen hundred (hours)
12.00	twelve hundred (hours)
21.00	twenty-one hundred (hours)
06.00	**oh** six hundred (hours) *or* **zero** six hundred (hours)
04.00	**oh** four hundred (hours) *or* **zero** four hundred (hours)

For minutes before or after the hour:

15.42	fifteen forty-two (hours)
09.05	zero (oh) nine five (hours)
07.20	zero (oh) seven twenty (hours)

UNIT 29: TIME (EXERCISES)

A) Write the following sentences using the long form for time. For example:

The moon will eclipse the sun at **5.00 am**.....**Five o'clock in the morning.**
The incubator must be switched off at **2.15**...**Quarter past two.**

1) The plane will be taking off at **7.15 pm**. ....
2) I logged on to the internet at **2.30**. ....
3) The automatic safe in the bank will open at **9 am**. ....
4) There will be a meeting for all the medical consultants at **4 pm**. ....
5) A Mondax pharmaceutical representative will arrive at **3**. ....
6) The project manager wants us in his office by **1 pm**. ....
7) The forecast is for rain by **9 pm**. ....
8) You have an interview with a scientific journalist at **4.30 pm**. ....
9) At **8.45 am** we will take the coach to the conference. ....

B) Choose the times from the box below that match the times in the sentences. For example:

Let's discuss your chemistry textbook over lunch at **one fifteen**...........**1.15**
The earthquake occurred at **six fifty-seven**..................................**6.57**

8.30	6.57	12.08	3.20	1.15	2.35	11.15	12.30	08.30

1) There will be a group seminar at **three twenty** every Friday. ....
2) You should stop heating the sample at **one fifteen**. ....
3) I will be giving a lecture on laboratory safety at **twelve thirty**. ....
4) You must be in London by six to catch the Southampton train at **six fifty-seven.** ....
5) Computing classes will be given at **eight thirty** every morning. ....
6) The electronic clock stopped at **two thirty-five**. ....
7) The professor phoned at **twelve oh eight** and wanted you to phone him at once. ....
8) The journalist from the British Medical Journal will be here at **eleven fifteen**. ....

C) Answer the following questions using the times in brackets. For example:

When should I meet you at the restaurant? (3, *short form*) **Meet me at three.**
When do you want me to arrive at the court? (16.00) **Arrive at sixteen hundred hours.**
When does the meeting start? (4, *long form*) **It starts at four o'clock.**

1) When did the pathologist arrive at the scene of the crime? (15.00)

....

2) If we leave now when will we arrive at the laboratory? (9.20, *short form*)

....

3) When did the storm start? (9.27, *short form*)

....

4) We can discuss the design of the experiment over lunch. When are you free? (1, *long form*)

....

5) When did the forensic scientist arrive at the court? (2, *long form*)

....

6) At what time does the research and development division close? (5.30, *long form*)

....

7) When will the DVD recorder switch on? (6.45, *short form*)

....

8) When does the lecture on radiation safety end? (11.30, *long form*)

....

UNIT 30: DATES

A) Writing dates

Dates can be written in two ways; long form and short form.
The long pattern is usually used at the top of letters and in most formal writing.
The short pattern is often used in application forms/questionnaires etc.

	SHORT FORM	LONG FORM
Written form	Day/Month/Year	Day/Name of month/Year
Example	**24/7/98**	**24th July 1998**
Reading and speaking written dates	"The twenty-fourth of the seventh, ninety-eight"	"The twenty-fourth of July nineteen ninety-eight"

A. Long form dates

1. In long form dates ordinal numbers are usually used.
For example:

1**st**	21**st**	31**st**
2**nd**	22**nd**	
3**rd**	23**rd**	

All other days are written with a "**-th**" For example:
4**th**, 5**th**, 6**th**, 11**th**, 12**th**, 13**th**

2. The "**-th**" etc., is sometimes omitted when writing the long form:
24 July 1998
1 September 1995

B. Short form dates

When using the short form, American English reverses the day and month:
U.K. **24/7**/11 <=> *U.S.A.* **7/24**/11

B) Dates in conversation

I. The fullest description of date is of the type:
Sunday **the** eleventh **of** July, nineteen ninety-three
Friday **the** twelfth **of** November, two thousand and twenty
Monday **the** thirteenth **of** October, two thousand and three
This is a formal description of the date as might be used, for example, in a court of law.

II. Informally the date may be said:
Sunday **the twelfth.** (The month and the year are clear from the context of the conversation).
The twelfth of July. (The day and the year are clear from the context of the discussion).
It is also possible to say *July twelfth* instead of the *twelfth of July*.

III. Prepositions used with the date;
When referring to a specific date, "**on**" is the preposition that is used. When a period in time is being referred to, then an "**in**" is used. For example:
Darwin published *The Origin of Species* **in** 1859.
The rainfall is greatest **in** April.
I arrived **on** the 24th July 1998.
She was born **on** the 3rd December 2007.

In American English "**on the**" can be omitted.

UNIT 30: DATES (EXERCISES)

A) Write down the following dates in words. For example:

31st July 2011 **The thirty-first of July two thousand and eleven**
6/8/92 (U.K.) **The sixth of the eighth ninety-two**

1) 10/11/93 (U.K.) ....
2) 1st April 1984 ....
3) 11th May 2001 ....
4) 3/5/97 (U.K.) ....
5) 22nd February 1953 ....
6) 24/1/98 (U.K.) ....
7) 17th November 1989 ....
8) 7th March 1995 ....
9) 31/12/72 (U.K.) ....
10) 24th June 1964 ....

B) Write down the following words as dates. For example:

The eighth of March two thousand and seven (long form)..........**8th March 2007**
The twentieth of the tenth eighty-two (short form)..........**20/10/82**

1) Twenty-third of August nineteen eighty (long form)
....
2) First of the third eighty-four (short form)
....
3) Seventeenth of September two thousand and ten (long form)
....
4) Twenty-ninth of February two thousand and twelve (long form)
....
5) Second of the third eighty-five (short form)
....

tenth	in	Friday	on	January	sixty-four

C) Complete the following sentences with words from the above table. For example:

"I'll meet you at the Dion II conference .....***in***.... March."
"The Lendrax sales representative will be here on the ....***tenth***... of August."

1) "There will be a test flight for the new plane on Tuesday the sixth of .... two thousand and thirteen."
2) I hope that sometime .... August, we will have a computer that can accurately read human writing.
3) "I can't see you on the ...., how about the fifteenth?
4) The technology was first used in .... but now it is out of date.
5) The organizer of the conference says that it will start on Monday and finish on ....
6) Are more children born .... a Saturday than on a Tuesday?

D) Answer the following questions using the dates given. For example:

When will the review article be published? **(7/8)**
"The review article will be published on the seventh of August."
When did you become a member of the Royal Physical Society? **(2nd March 1984)**
"I became a member of the Royal Physical Society on the second of March nineteen eighty-five."

1) When does the term (semester) start? **(4/10)**
....
2) When did Newton state the Law of Gravitation? **(1685)**
....
3) When will the plane be ready for use? **(July)**
....
4) When will we next see Halley's comet? **(1st January 2062)**
....
5) When can we expect the delivery of the new computers? **(12/9)**
....
6) When will the expedition to Kenya begin? **(17th August 2015)**
....
7) When was the last Californian earthquake? **(1st October 1993)**
....

UNIT 31: TIME AND DATES

The following periods of time are frequently referred to in spoken and written English:

Period	***Definition***	***Period***	***Definition***
Day	24 hours	Quarter	3 months
Weekend	Saturday and Sunday	Year	365 (or 366) days
Weekdays	Monday - Friday	Decade	10 years
Fortnight	14 days	Century	100 years
Month	28 – 31 days	Millennium	1000 years

For example:
When was Boyle's law discovered? **In** the 19th (*nineteenth*) century.
Personal computers became very popular **in** the 80s (*eighties*).
In the fourth millennium we will have to colonize another planet.
In April there were 6 inches of rainfall.

DURATION

A) Prepositions

For is used to describe the **length of time**:

The examination continued **for three hours**.
For two days, I waited for a reply.
I waited for a reply **for two days**.
The experiment lasted **for six hours**.

Over (F) and **during** (N) have similar meanings; they describe a period of time in which an *event occurs*.
"**During** the meeting I took a lot of notes but now I don't understand them."
"**Over** the meeting I took a lot of notes but now I don't understand them."
"**Over** the course of the experiment readings were taken every 10 minutes."

B) Verbs

Both *take* and *last (for)* refer to how long an activity occurs. **Take** is usually used when the activity is being performed by a person.

I took three hours to analyze the data.
It took **me** three hours to analyze the data.
Analyzing the data took **me** three hours.
A well maintained car can **last for** years.
The conference **lasted for** three days.

REPEATED PERIODS

C) Every

To refer to an event that occurs after a constant period we use "every":
"I took a reading of the patient's temperature **every four hours**."
"During the experiment I took a reading **every five minutes**."
"The sun comes up **every day**."
"Winter comes **every year**."

> every **day** = **dai**ly, every **week** = **week**ly, every **fortnight** = **fortnight**ly, every **month** = **month**ly, every **quarter** = **quarter**ly, every **year** = **year**ly, every **hour** = **hour**ly

It is possible to say:

I took temperature readings at hourly intervals.

This means the same as:

I took temperature readings hourly.

UNIT 31: TIME AND DATES (EXERCISES)

A) Answer the following questions using the periods given in brackets. For example:

When is the next meeting of the Mathematical Society? (January)...*The next meeting is in January.*
How often is the Journal published? (Quarter).......*The Journal is published every quarter (or quarterly).*

1) When do IBM plan to buy Apple? (2021)

....

2) How long will it take to grow the cells? (fortnight)

....

3) How long do you think this geological structure took to develop? (millennium)

....

4) When will the birds migrate south? (4 months)

....

5) When does the Voyager Spaceship land on Uranus? (next year)

....

6) How long will it be before we have a solution to the AIDS problem? (decade)

....

7) When will the drug tests be completed? (2 weeks)

....

8) When will the slides that I need for my presentation be ready? (2 days)

....

B) Choose one of the words from the box to complete the sentences below.

The words in the box may be used more than once.

During	**in**	**over**	**for**	**every**

For example:
He has a regular routine; ...**every**... Monday he starts an experiment.
The carbon dioxide produced by the yeast causes the bread to rise, this happens...**in**... an hour.

1) Did you take any notes .... the lecture or were you too busy listening?
2) How long do you have to centrifuge the sample ....?
3) Doctors often find tumours .... routine examinations.
4) .... the thunderstorm the wind speed reached 70 mph.
5) February had 29 days in 1992, 1988, 1984, so it's clear that February has 29 days .... four years.
6) I have bought a RISC based portable computer that will last me .... at least 5 years.
7) The effects of steroids last .... many months after you stop taking them.

C) Answer the following questions using the information in the brackets:

How often does the sun rise? *(every 24 hours)* The sun comes up daily.
How often do you check the air pollution in this city? *(every month)* I check the air pollution monthly.

1) How often must the plant be watered? *(every 24 hours)*

....

2) How often do you log onto the internet? *(every week)*

....

3) How often are you taking cocaine? *(two hours)*

....

4) How often do you have to apply for a grant? *(every year)*

....

5) How often is the centrifuge serviced? *(every five years)*

....

6) How often do you supervise practical work? *(every month)*

....

UNIT 32: MONEY

A) Notation

£ => Pound
$ => Dollar

p = pence or "p" *(pronounced as the letter "p" - "pee")*
c = cent

£ or $ always comes before the number to which it refers.
p always comes after the number that it refers to. For example:

£102.30p ⇨ One hundred and two pounds **and thirty** pence.
One hundred and two pounds **and thirty** p. *(Informal)*

When **completing a cheque** the whole sum number should be written in words. After the decimal point the phrase should begin with "**and**" and be completed with "**only**". The written number should always end with the word "**only**." It is good practice to prevent further figures from being added to the cheque by adding an ——— (*a horizontal line*) to the end of the figure written:

£167.52p ⇨ One hundred and sixty-seven pounds **and** fifty-two pence **only**———
£ 8.54p ⇨ Eight pounds **and** fifty-four pence **only**———
£2000.00 ⇨ Two thousand pounds **only**———

B) Exchange rates

If the exchange rate of *dollars to pounds* is **3:1**, then we can say:
The exchange rate is **three dollars to the pound**.
You can exchange **three dollars for every pound**.

If the exchange rate from *dollars to pounds* is **1:4**, then we can say:
The exchange rate is **one dollar to four pounds**.
You can exchange **one dollar for every four pounds**.

If there is a decimal in the exchange rate, then the decimal number is read or spoken. For example, if the exchange rate of *dollars to the pound* is **3.54:1**, then we can say:
The exchange rate is **three point five four dollars to the pound**.
You can exchange **three point five four dollars for every pound**.

Alternatively if the exchange rate of *dollars to the pound* is **1:3.54** then we can say:
The exchange rate is **one dollar to three point five four pounds**.
You can exchange **one dollar for every three point five four pounds**.

Informally, the exchange rates can also be read in a manner similar to mathematical expressions. For example **2.5 Yen = 1 Peso**, could be read as **two point five yen equals one peso**.

C) Common Currencies

Country	*Currency*	*Plural*	*Country*	*Currency*	*Plural*
Australia	**Dollar, A$**	**Dollars**	**Jersey**	**Pound, £**	**Pounds**
Brazilian	**Real, R$**	**Reais**	**Mexico**	**Peso, $**	**Pesos**
Brunei	**Dollar, B$**	**Dollars**	**New Zealand**	**Dollar, $**	**Dollars**
Canada	**Dollar, C$**	**Dollars**	**Pakistan**	**Rupee, Rs**	**Rupees**
Chile	**Peso, $**	**Pesos**	**Russia**	**Ruble, R**	**Rubles**
China	**Yuan, ¥**	**Yuan**	**Singapore**	**Dollar, $**	**Dollars**
Ghana	**Cedi, ¢**	**Cedis**	**Sri Lanka**	**Rupee, Rs**	**Rupees**
Hong Kong	**Dollar, $**	**Dollars**	**Sweden**	**Krona, kr**	**Kronor**
India	**Rupee, Rs**	**Rupees**	**Switzerland**	**Franc, F**	**Francs**
Iran	**Rial, Rls**	**Rials**	**United Arab Emirates**	**Dirham, DH**	**Dirhams**
Iraq	**Dinar, ID**	**Dinars**	**United Kingdom**	**Pound, £**	**Pounds**
Japan	**Yen, ¥**	**Yen**	**United States**	**Dollar, $**	**Dollars**

The European currency, the euro, is designated by an "€."

UNIT 32: MONEY (EXERCISES)

A) Write out the following exchange rates as words, in a similar manner to the examples below:

35 Rials : 1 Yen
The exchange rate is **thirty-five rials to the yen**.
200 Pesos : 1 Euro
The exchange rate is **two hundred pesos to the euro**.

1) If the exchange rate rises above **$5 : £1**, we will not be able to afford to go on holiday.

....

2) **34 Euros : 5 Yuan**

....

3) It is predicted that in two years the exchange rate will be **DH1 : Rs1**.

....

4) The advertised exchange rate was **¢100 : £1**, however, we received **¢200** for every pound.

....

5) If the euro is devalued then the pound to dollar exchange rate will rise above **£5 : $1**.

....

6) **10 Dinars : 1 Yen**

....

7) The ruble has become a more stable currency; its exchange rate of **R12 : £1** has been constant for a year.

....

B) Write out the following amounts of money as if they were being written in a cheque. For example:

£175.00
One hundred and seventy pounds only———
£35.17p
Thirty-five pounds and seventeen pence only———

1) $203.00

....

2) £85.00

....

3) R1000

....

4) $958.34

....

5) £76.50

....

6) Rs895

....

C) Change the following phrases or sentences, into the equivalent exchange rate. For example:

A thousand rupees equals 1 pound ⇨ **Rs1000 = £1**
An exchange rate of a thousand rupees to the pound ⇨ **Rs1000 : £1**

1) Two dollars equals one pound.

....

2) Six pesos for every two dollars.

....

3) Eight point seven five rubles for every pound.

....

4) You get twenty rupees for every dollar.

....

5) One pound gets you thirty-seven point four pesos.

....

UNIT 33: SINGULAR AND PLURAL

Most English plurals are generated by adding –s or –es to the singular form. Such examples would include *microscope/microscopes* and *virus/viruses*. However, there are a range of apparently irregular plurals that exist in scientific, medical and technical English that are usually derived from other languages. The commonest of these plurals are described below:

Table 1: The Basic Rules

Singular	Plural	Examples	Examples of Exceptions
-is	**-es**	Crisis Crises; See Table 2.	Patois Patois
-um	**-a**	Medium Media; See Table 3	Medias (exists in both forms) Museum Museums
-on	**-a**	Criterion Criteria	Electron Electrons Neuton Neutrons
-ex	**-ices**	Apex Apices	Reflex Reflexes
-a	**-ae**	Curriculum vita Curriculum vitae	Genitalia Genitalia Agenda Agendas
-ma	**-mata**	Dogma Dogmata	Dogmas (exists in both forms)
-en	**-ina**	Lumen Lumina	Lumens (exists in both forms)
-ix, -ax	**-ices** **-aces**	Helix Helices Thorax Thoraces	Helixes (exists in both forms) Matrix Matrices or Matrixes
-nx	**-nges**	Phalanx Phalanges	Lynx Lynx or Lynxes
-yx	**-yces**	Calyx calyces	---------------
-is	**-ides**	Vasculitis Vasculides	Crisis Crises; see above
-us	**-i**	Nucleus Nuclei Stimulus Stimuli	Genus Genera Corpus Corpora Apparatus Apparatus
-y	**-ies**	Deformity Deformities	July Julys

Table 2: These plurals are derived from the Ancient Greek language.

Singular	**Plural**
Basis	Bases
Hypothesis	Hypotheses
Thesis	Theses
Diagnosis	Diagnoses
Prognosis	Prognoses
Crisis	Crises
Diaphysis	Diaphyses
Metaphysis	Metaphyses
Epiphysis	Epiphyses
Dialysis	Dialyses
Analysis	Analyses
Axis	Axes
Ellipsis	Ellipses
Synthesis	Syntheses
Synopsis	Synopses
Metastasis	Metastases

Table 3: These plurals are derived from the Latin language.

Singular	**Plural**
Quantum	Quanta
Datum	Data
Phylum	Phyla
Bacterium	Bacteria
Ileum	Ilea
Medium	Media
Curriculum	Curricula
Diverticulum	Diverticula
Stratum	Strata
Erratum	Errata
Septum	Septa
Addendum	Addenda
Labium	Labia
Minimum	Minima
Ovum	Ova
Antrum	Antra

Because of the number and range of irregular plurals, a prudent student will always look up unfamiliar or unknown singular/plurals in a dictionary. Some plurals are so unique that they are best learned through experience; for example, spec**ies** (singular) and spec**ies** (plural).

UNIT 33: SINGULAR AND PLURAL (EXERCISES)

A) Can you convert each of the examples below into its correct plural?

1) Lamina	______________	8) Psychosis	______________
2) Larynx	______________	9) Bronchus	______________
3) Perineum	______________	10) Peritoneum	______________
4) Phenomenon	______________	11) Alga	______________
5) Locus	______________	12) Metropolis	______________
6) Meiosis	______________	13) Proton	______________
7) Syllabus	______________	14) Paralysis	______________

B) Can you convert each of the examples below into its correct singular?

1) Bacilli	______________	8) Strata	______________
2) Spectra	______________	9) Genera	______________
3) Abscissae	______________	10) Species	______________
4) Vortex	______________	11) Nebulae	______________
5) Foci	______________	12) Oases	______________
6) Emphases	______________	13) Neuroses	______________
7) Arthritides	______________	14) Radii	______________

C) Complete each of the sentences below with the suggested word. Use the correct singular or plural form, as appropriate:

How many _________ (*species*) of streptococci do you know?
How many **species** of streptococci do you know?

1) There are twelve thoracic ____________ (*vertebra*).

....

2) Calculate the ____________ (*velocity*) of the three projectiles.

....

3) We need to harvest three __________ (*ovum*) for in vitro fertilization.

....

4) ___________ (*Hypnoses*) is a recognized psychological therapy.

....

5) How many __________ (*proton*) were released as a result of each nuclear reaction?

....

6) Use the electron microscope to measure the length of each ______________ (*staphylococci*).

....

7) If we carried out a ___________ (*curricula*) review then we would have to change all of the ________ (*syllabus*).

....

8) Considering wave-particle duality, can you calculate the equivalent mass of a _______ (*photon*)?

....

9) These conjoined twins have two pubic __________ (symphysis).

....

10) The mathematician drew a ___________ (*matrices*) as part of his proof.

....

UNIT 34: APPROXIMATIONS

It is not always necessary to give accurate answers to numerical questions; approximations are often useful in discussions. This unit covers commonly used phrases.

A) Expressions used in approximations:

⇨ INCREASING ACCURACY ⇨		
More or less (N) Roughly (N) Pretty much (I)	Nearly (N) Approximately (N) About (N) Almost (N) Approaching (N) Just about (I) Going on for *(I) Getting on for* (I) Coming up to* (I)	Precisely (N) Exactly (N)

(N)=Neutral (I)=Informal

*These expressions have the same meaning as "approaching" or "nearing" but are less formal; they are more likely to be used in spoken English.

The approximation comes immediately before the number to which it refers.

For example:
The value of π is **roughly** 3
The value of π is **nearly** 3.15
The value of π is **precisely** 3.141593

There are **almost** 6 billion people on the planet.

B) Emphasis in approximations:

Very can be used before the following approximations to increase the strength of the approximation used:

⇨ INCREASING ACCURACY ⇨	
very roughly very approximately	very nearly

For example:
There are very roughly 5 billion people on the planet.
The farm covers an area of very nearly 10 square kilometers.

C) Similarly, the following phrases can be used to modify the strength of the approximation:

Well over (N): The number of people present at the meeting was well over 50.
Well under (N): The percentage experimental error was well under 10%.
Much more than (N): I can't remember the figure but it was much more than $5,000.
Much less than (N): His weight was much less than 80 kg.
Much greater than (N): The altitude was much greater than he could tolerate.
Far more (N): In 50 years the population of the country will be far more than 60 million.
I am treating far more than 20 patients per day.
Far less (N): Your percentage body fat must be far less than 40% to be healthy.
Just over (N): His height was just over 170 cm.
Just under (N): The hard disk can store just under 200 gigabytes of data.

UNIT 34: APPROXIMATIONS (EXERCISES)

A) Which of the following statements is more accurate? For example:

i) Her blood pressure was precisely 130/85 mmHg.
ii) Her blood pressure was more or less 130/85 mmHg.
i) is more accurate.

1) i) The global temperature will rise by about 2 degrees over the next 100 years.
ii) The global temperature will rise by roughly 2 degrees over the next 100 years.
2) i) The order of magnitude estimate for the distance to Mars is very roughly 400 million km.
ii) The order of magnitude estimate for the distance to Mars is very nearly 400 million km.
3) i) The resistance is nearly 100 ohms.
ii) The resistance is much less than 100 ohms.
4) i) The wind speed is precisely 30 miles per hour.
ii) The wind speed is approaching 30 miles per hour.
5) i) The number of species in this habitat is roughly 300.
ii) The number of species in this habitat is just about 300.
6) i) The density of the paper is pretty much 0.5g/100 sheets.
ii) The density of the paper is nearly 0.5g/100 sheets.
7) i) The height of this 100 storey building is exactly 2000 feet.
ii) The height of this 100 storey building is precisely 2000 feet.
8) i) The intelligence quotient (i.q.) of this group is very approximately 100-130.
ii) The intelligence quotient (i.q.) of this group is about 100-130.

B) Change the *italicized* words to another word or phrase with a similar meaning.

For example:
The density of steel is *nearly* the same as iron.
The density of steel is *almost* the same as iron.

1) Roughly how many of the cells are dead?
If you look down the microscope you can see that *about* 50% of the cells are dead.

....

2) Exactly how much acid will I need? *Precisely* 3.5 ml of acid will neutralize the alkali.

....

3) Will this be a successful personal organizer?
Yes, we expect to sell *approaching* 250,000 units in the first year.

....

4) The wavelength of visible light is *very roughly* 10^{-7} metres.

....

5) Approximately how often will I have to back up (= store) the data on the hard disk?
Roughly once a month should be okay.

....

6) How fast was the train traveling at that point? *Getting on for* 110 mph.

....

7) The latest population census in the UK showed the number of people was *approaching* 65 million.

....

8) Human Resources thinks that the company's employees may have to take a pay cut of *exactly* 10%.

....

9) Could you estimate the total number of students in this university? Yes, it's *getting on for* 15,000.

....

10) How much will 1000 shares in this company cost? £100,000 *more or less*.

....

UNIT 35: DIRECTION

A) Direction of movement:

-Wards or -ward are added to nouns and adverbs to indicate a **direction of movement**:

Towards A-------⇒-------⇒------⇒B The electron moved **towards** the positive charge.
Forwards A-------⇒-------⇒------⇒ Birds can only fly **forwards**; they can't fly backwards.
Backwards ⇐---------⇐--------⇐----A When you shoot, the gun will move **backwards**.

Upwards Hot air is less dense and so moves **upwards**.
Hot air is less dense and so moves up.

Downwards Gravity causes objects to fall **downwards**.
Gravity causes objects to fall down.

Similarly: She moved **in**wards.
He moved **out**wards.

> **Useful verbs are** attracted to **and** repelled by**. For example:**
> **The electron is** attracted to **the positive charge. So the electron moves** towards **the charge.**
> **The magnets with opposing polarities** are repelled**. So they move** away from **each other.**

B) Geographical directions:

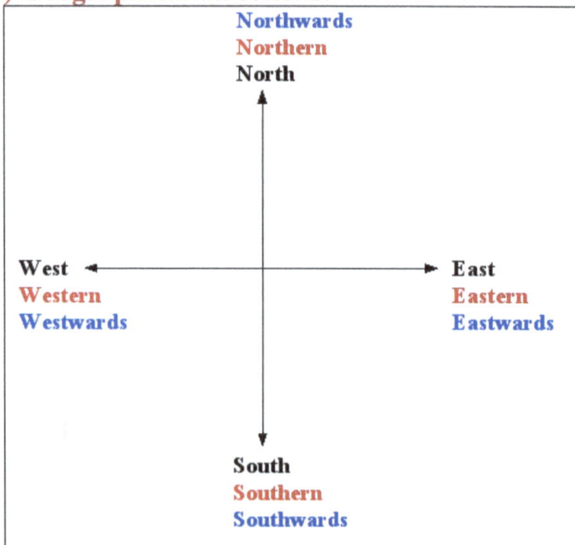

More complex directions can be described in the following ways:

The plane left by flying in a northeastern direction:

↗

It would return by flying in a southwestern direction:

↙

For example:

Northward(s) The wind was blowing northwards.
The wind was blowing north.
The wind was blowing in a northern direction.

Southward(s) The birds were flying southwards.
The birds were flying south.
The birds were flying in a southern direction.

C) Relative Positions:

Note these examples:

Where is Glasgow relative to London? Where is Glasgow in relation to London?
Glasgow **is north of** London.
How far is Glasgow from London?
Glasgow **is 403 miles north of** London.

UNIT 35: DIRECTION (EXERCISES)

A) Answer the following questions in a similar manner to the examples:

In what direction was the car moving? (**North**)
The car was moving north.
In what direction were the geese flying? (**Southwards**)
The geese were flying southwards.
In what direction is the rocket aimed? (**Eastern**)
The rocket is aimed in an eastern direction.

1) In what direction was the train moving? (**Western**)

....

2) Where is New York in relation to Los Angeles? (**East**)

....

3) How far is London from Accra? (**3175 miles, North**)

....

4) In what direction is the wind blowing? (**Southwestern**)

....

5) In what direction did the electron move? (**Towards, Electric field**)

....

6) From what direction is the hurricane coming? (**North**)

....

7) Where is the Radiology department in this hospital? (**East wing**)

....

B) Complete the following sentences with vocabulary from this unit. For example:

Is centripetal force directed inwards or _______?
Is centripetal force directed inwards or **outwards**?

1) Osmotic pressure occurs because water tends to move from its initial compartment ________ the higher solute concentration.

....

2) To reach the conference, you will have to drive south for 50 miles and enter the city centre. The conference venue is 100 metres north ______ the hospital.

....

....

3) The earth rotates around the sun because of a gravitational force ________ the Earth to the Sun.

....

4) The explosion propelled fragments ___________ for 100 metres.

....

5) Your compass should always point ___________ magnetic north.

....

6) The Maglev trains work because the trains are _______ by the tracks and so float.

....

7) In what direction is the ship sailing? The ship is sailing in a southwestern __________.

....

8) Blood flows ______ from the heart into the aorta and then _______ to the iliac vessels.

....

UNIT 36: MOVEMENT

A) Circular movement

The verbs **spin** and **rotate** have similar meanings and can be used interchangeably:

Spin (N) The planet Earth **spins** on its axis with an approximately 24 hour period.
The planet Earth **rotates** on its axis with an approximately 24 hour period.

Rotate (N) If you examine the film correctly you can see that the ballerina **rotates** exactly three times.
If you examine the film correctly you can see that the ballerina **spins** exactly three times.

Revolve (N) Revolve tends to be used to describe the movement of objects and not people. Also remember that an **object revolves,** it is not revolved:
When you spun the coin, it **revolved** 30 times before it came to rest.
Mars **revolves** around the Sun.

Turn (N)
Twist (I) **Turn** and **twist** have similar meanings and can be used interchangeably; however, **twist** is a less formal word.
He **turned** the key in the lock.
He **turned** just in time to see the cricket ball coming towards him.
She **twisted** open the jar of peanut butter.
During the 100 metres race he fell over and **twisted** his ankle.

To circle (N) The shark **circled** its prey before attacking.
The plane **circled** the airport once before landing.

To describe a circle (F)
When this subatomic particle enters an electric field, it will **describe a** perfect **circle**.

Note this alternative construction; a verb of movement + *in a circle*. For example:
He was a nervous public speaker who walked in a circle **as he spoke.**

B) Linear movement

Here, it is common to use a verb of motion followed by **in a straight line**:

After the snooker ball was struck, it moved **in a straight line**.
The collision caused the proton to travel **in a straight line** away from the impact site.
In the final part of the mating ritual, the male walks towards the female **in a straight line**.

Other variations include - straight up, straight down, straight across, straight through etc:
"After the bullet was fired it passed **straight through** the skin and soft tissues and entered the aorta."

C) Random movement

The commonest constructions involve a verb of movement followed by ***randomly***. For example:
Brownian motion is one type of random motion; an object undergoing Brownian motion **moves randomly**.
You should not **wander randomly** through this area because it is a minefield.

D) Collisions

Collide with (N) and **hit** (I) can be used interchangeably to indicate collisions.
The purpose of a cyclotron is to accelerate particles so that they can **collide with** the target.
The purpose of a cyclotron is to accelerate particles so that they can **hit** the target.
A professional boxer can throw a punch that **collides with** a face with a pressure of more than 1500 pounds per square inch.

(I) = Informal, (N) = Neutral, (F) = Formal

UNIT 36: MOVEMENT (EXERCISES)

A) Rewrite the following sentences as the appropriate questions, in a manner similar to these examples:

If the car continues **in a straight line** it will **collide** with the van.
What will happen if the car continues in a straight line?
It takes 365 days for the Earth **to rotate** around the sun.
How long does it take for the Earth to rotate around the sun?
Because of its trajectory it **collided with** the larger object.
Why did it collide with the larger object?

1) If you turn the dimmer switch clockwise you increase the brightness of the lights.

....

2) He can only turn his head by +/- 30 degrees because of a road traffic accident that damaged his cervical spine.

....

3) The Earth rotates seven times in a week.

....

4) He twisted his back whilst playing football.

....

5) If you keep driving in a straight line you will reach Paris.

....

6) A bee does not move randomly from one flower to another.

....

7) Circling your answer means that you have confirmed your selection.

....

8) Flying in a straight line, it takes three hours to reach London.

....

B) Complete the following sentences with vocabulary from this unit.

For example:
Is a _______ door more likely to cause accidents than a traditional door?
Is a revolving door more likely to cause accidents than a traditional door?

1) He fired the rifle from a distance of 1000 metres and yet he still _______ the centre of the target.

....

2) Does light travel in a _________ line?

....

3) The bull _______ with the matador and killed him.

....

4) Does an electron rotating around a nucleus _______ a circle?

....

5) The centrifuge is _______ at a rate of 2000 rpm.

....

6) At first sight the ants look as though they are moving _______, but if you watch carefully you can see that they have a purpose.

....

7) This hard disk ______ at 7200 rpm, however I think that I would prefer a fast flash drive.

....

8) The faster the spines of the windmill rotate, the greater the amount of electricity that the windmill produces. There is no wind now so the spines are not _______.

....

....

UNIT 37: COMPARISONS

A) Qualitative Comparisons

When describing two quantities, a comparative adjective may be used for a qualitative comparison. For example:

"The population of Japan is **larger than** the population of Britain." *Or*
"The population of Britain is **less than** the population of Japan."

An indication of the size of the difference can be given by using one of the following phrases:

Large Differences	Intermediate Differences	Small Differences
Considerably (**F**)	Somewhat (**F**)	Slightly (**N**)
Substantially (**F**)	Noticeably (**N**)	A little (**N**)
A great deal (**N**)	Distinctly (**N**)	Marginally (**F**)
Dramatically (**N**)		
Far (**N**)		
Much (**I**)		

"The population of Japan is **considerably greater than** the population of Britain."
"The population of Britain is **substantially less than** the population of Japan."
"Patients with acromegaly are **much taller than** they should be."
"The new high speed train is **a great deal faster than** the older train."

B) Quantitative Comparisons

When making more quantitative comparisons the phrases below may be useful.

As much as
As many as
As little as
As tall as
As fast as etc.

For example: "Product A earns twice **as much** money for the company **as** product B."
"The new car is half **as fast as** the old one and needs a third as much petroleum."
"Some one with osteogenesis imperfecta is three times **as likely** to have a bone fracture **as** a normal individual."

Percentages can be used to make truly quantitative comparisons:
"This metal has **20% of** the electrical resistance of copper!"
"The Moon has **60% of** the gravity of Earth."
"Why do children in this region have IQs **10% above** the national average?" *Or*
"Why do children in this region have IQs **110% of** the national average?"
"The bone density of this 70 year old woman is **15% of** the expected value."

C) Percentages and comparative adjectives

The comparative adjectives used in **(A)** can be combined with percentages used in **(B)**.
For example:

"The density of this metal is **30% less than** that alloy."
"The nitrogen dioxide air pollution in Ethiopia is **20% less than** in Canada."
"The nitrate content of the water is **40% greater** in the city **than** in the countryside."
"The Northern Europeans will be **15% taller than** the Southern Europeans in 100 years."
"This post surgical patient has a blood pressure that is **30% less than** I would have expected."

(I) = Informal, (N) = Neutral, (F) = Formal

UNIT 37: COMPARISONS (EXERCISES)

A) Answer the questions below with the information and vocabulary given. For example:

How does the top speed of this train compare with the previous one? **Greater, 15%**
The top speed of this train is **15% greater than** the previous one.
Who is more likely to develop colorectal cancer a vegetarian or a non-vegetarian? **Non-vegetarian, Much**
A non-vegetarian is **much more** likely to develop colorectal cancer.

1) Is this television show less popular with social class 1 or social class 2? **Social class 1, Marginally**

....

2) Does copper have a greater thermal conductivity than aluminium? **Larger, 40%**

....

3) Is his blood pressure higher than the average blood pressure? **No, 85% of**

....

4) Did the modifications to the engine make the car faster? **Faster, Noticeably**

....

5) Did the addition of the enzyme make reaction A faster than reaction B? **Yes, 25%**

....

6) How does the performance of this group of students compare with the last group? **Better, Considerably**

....

7) Are patients with achondroplasia smaller than average? **Yes, Far**

....

8) Is the probability of winning the National lottery equal to that of being struck by lightning? **Yes, as likely**

....

B) Complete the following sentences with vocabulary from this unit. For example:

Athletes take steroids because they think that they will perform much better with them _____ without them.
Athletes take steroids because they think that they will perform much better with them **than** without them.

1) Species A seems bigger than species B. In fact, if you compare the average heights of the two species you can see that, species A is significantly ______ than species B. This confirms my first impression.

....

2) I was hoping for definitive results from this experiment. However the data are not ______ conclusive _____ I expected.

....

3) The patient's urine output should be about 25 millilitres per hour. In the last hour it was only 5 millilitres.
This is far ______ than expected.

....

4) If you increase the speed of the processor and increase the size of the LCD screen, you will need a 100% greater power supply. My calculations confirm that the rechargeable battery will need to produce a power output that is _____ as large as the previous model.

....

5) A diet low in fresh fruit and vegetables greatly increases the chance of a range of cancers. For example, the probability of developing rectal carcinoma is 50% more in a low fruit and vegetable diet _____ in a high fruit and vegetable diet.

....

6) The maximum velocity of this train is only 85% ____ the predicted value.

....

7) To get a safety certificate, the material needs to be sufficiently flame retardant. Can this synthetic material resist flames _______ much as is required?

....

UNIT 38: PROCESSES

Any description of a physical, biological or chemical process should have a **beginning,** a **middle** and an **end**. There are verbs that can be used to describe each part of the process:

A) BEGINNING

Start (N) = Commence, begin. This is the most commonly used verb. This word can be followed by an infinitive or a gerund (-ing). For example:

The pharmaceutical company has **started to sell** a new antidepressant.
The pharmaceutical company has **started selling** a new antidepressant.
The reaction **starts** when the chemicals are mixed.
The engine is **started** by turning the switch.

Begin (N) = Start. It is followed by an infinitive or a gerund (-ing). Often the gerund requires the preposition ***by***. For example:

The airplane **began** *to accelerate* and 30 seconds later it left the ground.
I **began** by *describing* my invention.
I **began** to describe my invention.
Digestion of carbohydrate **begins** in the mouth.

Trigger (N) = Begin, start. **Trigger** is often used to refer to processes that are automatic or mechanical. This verb is often used in the passive:

The chemical reaction **is triggered by** light.
What **triggered** the explosion?
We are interested in what **triggered** the insanity.

Initiate (F) = Start, trigger. This verb is also used formally.

To **initiate** the process you must press this button.
To **initiate** the program you must type in the correct code.

B) PREPOSITIONS

Each of the verbs may be followed by a preposition:

At	**=> Used for Place or Time**
On	**=> Used for Place or Time**
In	**=> Used for Place**
With	**=> Used for Sequencing (first in a sequence)**
By	**=> Used for Cause/Reason**

For example:
The car started **at** 6 pm. (**At** for **Time**)
The molecule begins **at** the cell surface and then moves into the cell. (**At** for **Position**)

The rain began **in** the north of the country. (**In** for **Position**)

The sulphuric acid production process was initiated **on** Monday. (**On** for **Time**)
The fertiliser started **on** top of the hill but the rain washed it down. (**On** for **Place**)

Photosynthesis is initiated **by** light. (**By** for **Cause**)
Photosynthesis begins **with** the absorption of carbon dioxide and water. (**With** for **first** in sequence.)

(I) = *Informal*
(N) = *Neutral*
(F) = *Formal*

UNIT 38: PROCESSES (EXERCISES)

A) Complete the following sentences with the appropriate verbs and prepositions. For example:

The production of the televisions _____ __ 8am by switching on the power supply to the machinery.
The production of the televisions **began at** 8am by switching on the power supply to the machinery.

1) Respiration _______ _______ glycolysis and then continues with the Kreb's Cycle.

....

2) This chemical reaction is initiated ____ a photon of light.

....

3) The movement of this engine is _________ ________ the combustion of petroleum.

....

4) I started my presentation promptly _______ 1pm.

....

5) The morphine pump is __________ every time the patient presses the button.

....

6) The emotion of anger begins ______ the brain with a release of noradrenaline. However, it is not clear what _______ this release.

....

7) A biomechanical analysis of this tennis player's serve, _______ ______ the top of the swing.

....

B) Rewrite the following sentences as the most appropriate questions. For example:

Oxidative phosphorylation starts in the mitochondrion.
Where does oxidative phosphorylation **start**?
The chemical reaction is initiated by raising the temperature.
What initiates the chemical reaction?
The train began its journey at 11pm.
When did the train **begin** its journey?

1) The pharmaceutical company started to sell the new painkiller in August.

....

2) The sea breeze triggered rainfall over the east coast of India.

....

3) Coagulation starts because the temperature of the egg is raised.

....

4) The chemicals act as a nidus to initiate the crystallization of a urinary calculus.

....

5) Pepsin begins the digestion of protein in the stomach.

....

6) The download of this file was triggered by the user at 8.15pm.

....

7) The protocol starts with the dissolution of salt to make a 1M solution of sodium chloride.

....

8) Her journey around the world began in Afghanistan.

....

9) He started his proof by making assumptions that were not justified.

....

10) A smile begins with the contraction of the orbicularis oris muscle.

....

UNIT 39: PROCESSES II

The following are words and phrases used to describe processes that have already started, are continuing or are being completed.

A) MIDDLE

Maintain (N) This verb means that the process remains constant or unchanged and so is at the same level. **Maintain** is usually followed by a noun or equivalent phrase. If a specific level of maintenance is indicated then the preposition **at** is often used:
"The pilot **maintained** the **speed** of the plane **at** 700 miles per hour."
"To **maintain** the chemical reaction, there must be a continuous supply of reactants."
"This species first populated the area 20 years ago. However, in recent years it has been difficult to **maintain** its population because of the arrival of a new predator."
"We gave this critically ill patient an infusion of positive ionotropes to **maintain** his blood pressure."

Continue (N) This verb means that a process that has already started, ***carries on***. Continue can be followed by an infinitive (to -) or gerund (-ing). For example:
"The adrenal gland **continues** secret**ing** noradrenaline as long as the "fight or flight signal is received."
"Can Microsoft **continue to** sell the GLACIAL operating system?"
"Do cigarette smokers **continue** smok**ing** because they need nicotine?"
"After the nuclear reaction is initiated by proton collision, it **continues** by releas**ing** beta particles."

B) END

Finish (N) This verb means the same as *to end* or *to complete*. It is most commonly followed by a gerund; so you usually **finish doing something**. Finish can also be followed by a noun to indicate completion of a task or activity. For example:
"**Finish** kneading the dough and put it in the oven; the heat will denature the proteins in the flour.
"The download from that website finally **finished** 15 minutes after it started."
"After you **finish** dissolving the copper sulphate in water, leave it to gradually cool to room temperature."
"She **finished** the last lap 10 seconds ahead of her closest rival."

End (N) This verb means the same as *to finish* or *to complete* a process. It is usually followed by the noun to which it refers:
"We **ended** the examination of the new car with a crash test."
"The conference **ended** at 7pm because it could not be extended any further."

Stop (N) To halt a process. This is usually followed by a gerund (-ing). So again, you would **stop doing something**. Alternatively, it can be followed by the noun to which it refers:
"How long does it take to **stop** a centrifuge that is spinning at 30,000 rpm?"
"Despite giving the patient azathioprine, the rejection of the transplanted lung did not **stop**."
"As soon as I became aware of the allergy, I **stopped** prescribing penicillin."

Complete (F) This means the same as finish but is more likely to be used in a formal context. It is usually followed by a noun or equivalent phrase. For example:
"I **completed** the article on the 21/11/20 and it was published on 5/5/21."
"Oxidative phosphorylation is **completed** when the electron arrives at its final acceptor; O_2."

(N) = Neutral, (F) = Formal

The prepositions **at**, **on**, **by**, **with** and **in**, can be used with these verbs as (see previous unit).

UNIT 39: PROCESSES II (EXERCISES)

A) Complete the following sentences with the appropriate verbs and prepositions. For example:

The pipes ________ a water pressure of about 60 N/m^2 for most of their length.
The pipes **maintain** a water pressure of about 60 N/m^2 for most of their length.

1) Initially, the bullet is accelerated to a muzzle velocity of 975 m/s and then ________ a velocity of over 200 m/s for the first 100 m.

....

2) The crack in the crystal begins at point A, then _________ to point B and finally ends at point C.

....

3) I ______ reading the article and ordered one of the references. Then I _______ reading the article.

....

4) The thermostat will ensure that the waterbath is ________ at 37 °C. However, after an hour the reaction will be stopped by turning off the thermostat and by adding 2 mM NaCl.

....

....

5) At 10 pm I will _____ working on this presentation. However, I don't just want the work to be halted, I want the presentation to be properly ________.

....

....

6) If the plane _________ its current speed, it will arrive at Toronto airport in 3 hours.

....

7) If the plane _________ to fly in this direction, it will arrive at Toronto airport in 3 hours.

....

B) Complete the following sentences with verbs and prepositions from this unit or the previous unit.

1) The metabolism of bilirubin begins ___ the liver, _________ through the length of the small intestine and is completed _____ the large intestine.

....

....

2) *Formal scenario:*
Before you can start a building project of this magnitude, you must get funding. Then you must make sure that the project begins on time, ________ within its financial limits and is _______ promptly.

....

....

3) Cancer formation in the lower oesophagus begins ______ metaplasia, then ________ with dysplasia and ends _______ an invasive tumour.

....

....

4) *Formal scenario:*
First the reactants are adsorbed onto the surface of the catalyst. Then the reaction continues as one of the reactant achieves a high energy state whilst the other reactant ________ its original state.The proximity of the two reactants allows the reaction to occur. Finally, the reaction is _________ and the products are released.

....

....

....

5) This presentation must begin at 9 am and ________ at 10 am. The next speaker ________ at 10.05 am.

....

....

UNIT 40: PROBABILITY

A) Expressing probability in words:

⇨ ⇨ ⇨ ⇨ **Increasing Probability** ⇨ ⇨ ⇨ ⇨				
Impossible *(probability of zero)*	Improbable Unlikely	Possible	Probable Likely	Certain Definite Sure ***(probability of one)***

For example:
Are you **certain** that the molecular weight of water is 17?
The hypothesis you have suggested seems **possible** but in reality I think it is **unlikely** to be the true answer.
Which is more **probable**, that the patient is suffering from two common diseases or one rare disease?
According to the Heisenberg uncertainty principle it is **impossible** to simultaneously know the position and velocity of an electron.
How **likely** were you to have got an *A* on this assignment without cheating?

B) Comparing probabilities

A common construction involves using a comparative adjective before one of the words in the above table:
Is it **more likely** that a patient will die of cancer or heart disease?
It is **less likely** that he will win money at the casino than lose money.
It is **more probable** that a Ferrari will win the race than a Mercedes.
I am **more certain** that I want to emigrate to the USA than that I want to emigrate to Canada.

For probabilities that are equal, the following phrases can be used:

As likely as	**Equally likely**	**An equal likelihood**
As probable as	**Equally probable**	**An equal probability**
As certain as	**Equally certain**	**An equal certainty**
As sure as	**Equally sure**	------------------------
As uncertain as	**Equally uncertain**	**An equal uncertainty**
As definite as	**Equally definite**	------------------------
As possible as	**Equally possible**	**An equal possibility**
As improbable as	**Equally improbable**	**An equal improbability**
As unlikely as	**Equally unlikely**	------------------------
As impossible as	**Equally impossible**	**An equal impossibility**

For example:
If you toss a coin, you are **as likely** to get heads **as** tails.
If you toss a coin, there is an **equal likelihood** that you will get heads or tails.
If you shuffle the cards, then each player is **equally likely** to get a King, a Queen or a Jack.
If you shuffle the cards, then each player is **as likely as** the other to get a King, a Queen or a Jack.
Considering the current evidence, the two hypotheses are **equally possible**.
Considering the current evidence, the two hypotheses are **as likely as** each other.

C) Qualifying probabilities

An impression of the size of the probability can be indicated by using the following words:

It is **very** likely that the experiment will be successful.
It is **much more** likely that the new design will have more efficient aerodynamics.
The **most** probable outcome of increasing the density of the alloy will be to increase its strength.
It is **much less** likely that it will snow in the summer than in the winter.
You are **much more** likely to have hayfever in the summer than the winter.

UNIT 40: PROBABILITY (EXERCISES)

A) For each of the scenarios below, complete the sentences with the appropriate vocabulary:

Scenario 1) A statistician talking to a client.
Statistician: I can offer a range of services on a private basis.
Client: That sounds good. What I really need help with, is preparing my statistical data for publication.
Statistician: No problem. I can calculate the statistical significance of your results; in other words how ________ you were to have achieved the same results by random chance.
Client: Excellent. I am ________ that you will give me the help I need.

Scenario 2) A chemistry lecturer talking to an undergraduate student.
Student: I was interested in your description of the factors that control the reaction rate.
Lecturer: The most important consideration is the proportion of molecules that are above the activation energy for the reaction. Basically, the greater the proportion of such high energy molecules, the more ________ it is that the reaction will occur.
Student: I think I understand. You are saying that high energy molecules are more _________ to react.
Lecturer: That is essentially correct. It also follows that if two molecules have the same energy, the probability that the reaction will occur is equal.

Scenario 3) A zoologist giving a commentary to a television program.
"The mating ritual of these animals is prolonged and intricate. At first, the female looks upon the male as an unlikely or _________ mate. However, after two hours, she sees him as a possible mate. After another hour she sees him as a probable or __________ mate. If the ritual continues for four hours then the suitor becomes a sure or ________ mate."

Scenario 4) A doctor talking to a patient about the risk of lung cancer.
Doctor: I cannot give you an absolute measurement of your chances of getting lung cancer, but I can tell you some of the factors that increase the risk.
Patient: What factors?
Doctor: Smoking for example. It makes the _________ of getting lung cancer much greater.
Patient: Will I be sure to get lung cancer if I smoke?
Doctor: No. But the chances will go from an improbability to a possibility.
Patient: What is the best way to be _____?
Doctor: The only way to be certain is to perform a lung biopsy.

Scenario 5) Two doctoral scientists talking about a colleague during a coffee break.
Scientist 1: He has applied for a prestigious research grant and fellowship.
Scientist 2: Yes, I have heard that he is ________ that he will succeed. However with a **10/1** applicant to appointment ratio, I feel that he is being optimistic.
Scientist 1: The numbers suggest that it is possible that he will succeed. However, if you look at the figures objectively, it is _________ that he will be successful.
Scientist 2: What if he has the support of the Professor?
Scientist 1: That would change everything.

Scenario 6) An actuary explaining his job.
First of all, you have to understand that insurance companies are businesses that exist to make a profit. They must set their charges and rates at a level that is _______ to make money. Actuaries have the mathematical skills in probability and statistics to ensure that this is the case. For example, if you want to sell a 25 year old middle class male life insurance, you must calculate the probability of him dying in the next 40 years. Hopefully, the chances of this happening are small. It should be an _______ event but a good actuary can make a precise calculation of the expectation.

UNIT 41: TEMPERATURE AND TEMPERATURE CHANGES

A) There are three temperature scales commonly in use:

Centigrade or Celsius: Water boils at 100°C and freezes at 0°C.
Kelvin: Water boils at 212 K and freezes at 32 K.
Fahrenheit: Water boils at 373.15°F and freezes at 32°F.

°C is read as *degrees* centigrade.
K is read as kelvin.
°F is read as *degrees* fahrenheit.

B) Temperature

The temperature of an object can be described by the following adjectives:

⇒⇒ INCREASING ⇒⇒ TEMPERATURE ⇒⇒

Frozen	Cold	Cool	Lukewarm	Warm	Hot	Boiling

For example;

"Much of the water in the arctic is frozen."
"The patient must be kept cool."
"A boiling liquid becomes a gas."
"If you leave your cup of coffee for too long, it will go from hot to lukewarm."

C) To change the temperature of an object, the following verbs may be used:

Increasing the temperature -	to warm	↑	(*Informally "to warm up"*)
	to heat	↑↑	(*Informally "to heat up"*)
	to burn	↑↑↑	
Decreasing the temperature -	to cool	↓	(*Informally "to cool down"*)
	to chill	↓↓	
	to freeze	↓↓↓	

For example:

"Polythene burns if heated strongly."
"Is the world heating up?"
"If you lower the temperature in the fridge, the water will freeze."

> **All of the verbs are regular except** freeze **(freeze, froze, frozen).** Burn **has both regular and irregular forms; burn, burned, burned or *burn, burnt, burnt.***

D) The following verbs are used to describe changes of state (solid, liquid or gas):

(s) = solid (l) = liquid (g) = gas

(s)->(l)	(l)->(s)	(l)->(g)	(g)->(l)	(s)->(g)	(g)->(s)
melt liquefy thaw*	solidify	vaporise	condense liquefy	sublime	condense solidify

* Usually applies to water

For example:

"Iodine crystals sublime when heated."
"When water vapour condenses it becomes steam."

UNIT 41: TEMPERATURE AND TEMPERATURE CHANGES (EXERCISES)

A) Complete the following sentences with the most appropriate verb of temperature change.

For example: Did you know that the materials produced by ______ polymers are carcinogens?
Did you know that the materials produced by **burning** polymers are carcinogens?

1) The protocol states that the solution should be heated to 100°C for 30 minutes.

…………………………………………………………………………………………………

2) Dry ice (frozen carbon dioxide) can _______ from solid to liquid when _______.

…………………………………………………………………………………………………

3) __________ the organic solvent from gas to liquid is the final stage in the production process.

…………………………………………………………………………………………………

4) The water on the surface of the building ______ in very cold weather. Then the ice expands and damages the bricks.

…………………………………………………………………………………………………

5) This hypothermic patient must be ______ gradually to a normal body temperature.

…………………………………………………………………………………………………

6) This purification process relies on _________ the gold until it melts.

…………………………………………………………………………………………………

7) Did you know that when bears hibernate in the winter their core body temperatures _______ to 25°C?

…………………………………………………………………………………………………

B) Rephrase the following sentences as the most appropriate questions.

For example: Water boils at 100°C.
At what temperature does water boil?
This liquid becomes a solid at 50 K.
When does this liquid become a solid?
If warmed this solid becomes a gas.
What happens if this solid is warmed?

1) If the beverages are lukewarm then sales are poor.

…………………………………………………………………………………………………

2) Snow is formed as water falls through cold air and freezes solid.

…………………………………………………………………………………………………

3) After the thermostat was switched off at 10pm, the room cooled by 5°C.

…………………………………………………………………………………………………

4) The extensive burns on this individual were caused by boiling water.

…………………………………………………………………………………………………

5) Calcium salts burn in a flame with a brick red colour.

…………………………………………………………………………………………………

6) If the engine becomes too hot then the lubricating oil stops being effective.

…………………………………………………………………………………………………

7) Perspiration cools down the body by evaporation.

…………………………………………………………………………………………………

8) As the albumin solution is heated it becomes more opaque.

…………………………………………………………………………………………………

9) Bacteria proliferate more slowly in cool liquids than warm liquids.

…………………………………………………………………………………………………

10) This metal becomes a superconductor at 30 K.

…………………………………………………………………………………………………

11) If you increase the temperature of this biological system by 10°C, the reaction rate doubles.

…………………………………………………………………………………………………

12) After the plane crash the burnt bodies could only be identified by dental records.

…………………………………………………………………………………………………

UNIT 42: PRESENTATIONS

At some point it will be necessary to present information to colleagues in the form of a lecture, seminar, talk or presentation. The core components of such presentations are (i) **Introductory** phrases (ii) **Sequencing** phrases (iii) Reference to **diagrams/tables/graphs** etc., and (iv) **Summarizing** or concluding phrases.

(I) = Informal, (N) = Neutral, (F) = Formal

A) INTRODUCTION

The following phrases can be used to introduce a presentation:

I'm **going to talk** about.....(N)
I **will talk** about......(N)
The **aim of** this presentation......(F)
The **purpose of**....(N)

An introduction to a presentation may be made more formal by using the conditional tense:

I **would** like to begin this presentation with.....(F)

More informal (I) introductions can commence as shown:

Right then.....(I)
Okay, I am going to talk....(I)
So let's begin with......(I)

The more formal introductory phrases could be used at an external conference, for example. On the other hand, the less formal phrases could be used within a local department.

B) SEQUENCING

The sequencing words can be taken from either column below; however, no single sequence should include words from both columns:

Firstly (N),	First (N)
Secondly (N),	Second (N)
Thirdly (N),	Third (N)
Fourthly (N),	Fourth (N)
Fifthly (N),	Fifth (N)
Sixthly (N),	Sixth (N)
Seventhly (N) etc.	Seventh (N) etc.

"There are several factors that must be considered in choosing an MRI machine; **first,** there is the cost, **second,** there is the question of maintenance and **third,** a space must be found for it in the laboratory."

"**Firstly,** I am not convinced that funding will be easy to obtain. **Secondly,** I do not believe that this line of research will be productive in the short term. **Thirdly,** we may meet with some resistance from the general public."

"The **first point** I'd like to make.....The **second point** I'd like to make.....The **third point** I'd like to make......"

First.........second........**thirdly**..... *This is a common type of sequencing error.*

There are other simple phrases that are frequently used in sequencing:
Afterwards/After this/After that, we will move to the next subject....(N)
Next, I would like to talk about.......(N)
Then there is the problem of.......(N)
Now, I would like to move on to.....(F)

UNIT 42: PRESENTATIONS (EXERCISES)

A) Complete the following sentences with vocabulary from this unit.

Formal scenario:
Good morning, ladies and gentlemen. I _______ like to start this presentation by referring to this slide.
Good morning, ladies and gentlemen. I **would** like to start this presentation by referring to this slide.

1) Having discussed the costs of the suggested changes, I would now like to move _____ _____ a consideration of the benefits.

....

2) So, let's begin _____ a description of the normal anatomy of the human shoulder. Afterwards, we can consider some of the common shoulder pathology.

....

....

3) Firstly, the current form of the periodic table will be discussed. Secondly, the solid non-metals will be examined. _________, the gaseous non-metals will be examined. __________ and finally, the rest of the elements will be considered.

....

....

4) I'm going to _____ about the causes, mechanisms and risk factors for diabetes mellitus.

....

5) The _____ of this presentation is to review the latest developments in subatomic physics.

....

B) Are the following sentences correct examples of introductory phrases/sequencing phrases?

For example:
I shall like to begin this presentation with..............☒ (Incorrect)
I would like to begin this presentation with.............☑ (Correct)

1) Right then, I'm going to talk about North American economics over the last decade.
☐ (Incorrect) ☐ (Correct)
2) Firstly, we will address the regional politics, secondly, the regional economics and third, the regional demographics.
☐ (Incorrect) ☐ (Correct)
3) We will talk about the design of our water-cooled supercomputer.
☐ (Incorrect) ☐ (Correct)
4) Since then, the purpose of this presentation is to review recent research.
☐ (Incorrect) ☐ (Correct)
5) So, let's initiate with a discussion of the increased risk of heart attacks in migraine sufferers.
☐ (Incorrect) ☐ (Correct)
6) Now, I would like to move on to a consideration of statistical significance testing.
☐ (Incorrect) ☐ (Correct)
7) So, let's begin with the script writing, then the direction, after that the background music and finally the acting that this film demonstrates.
☐ (Incorrect) ☐ (Correct)
8) I will talk for about 10 minutes on wave particle duality and its implications.
☐ (Incorrect) ☐ (Correct)
9) The aim in this talk is to review progress in alternative energy sources.
☐ (Incorrect) ☐ (Correct)
10) Right then, we will start with a description and discussion of solid state nuclear magnetic resonance.
☐ (Incorrect) ☐ (Correct)
11) First, we will discuss the pathology of asthma and then we will consider its management.
☐ (Incorrect) ☐ (Correct)

UNIT 43: PRESENTATIONS II

A) A reference to graphs, tables or other visual aids can be introduced by:

Turning to table VI ....(N)
Looking at figure 7b....(N)
The **graph/table/figure shows**.....(N)
As you can see on the chart....(N)

If the speaker wishes to increase the level of formality, "*Ladies and gentlemen.....*" can be used to introduce any of the above the phrases:
"**Ladies and gentlemen**, **turning to the chart** it is clear that we have steadily increased the number of publications per year, over the last decade."

Another technique for increasing the level of formality of the basic expressions above makes use of a conditional phrase:

"**If I could ask you to** (turn to/look at the graph). It shows a plot of temperature against time...." (F)
"**Could I ask you to** look at figure H....." (F)
"**Would you now turn to**......." (F)

B) Summary/Conclusion

Although many different verbs can be used to conclude a presentation, they can all be used in similar ways:

In conclusion......(N) To conclude.......(N) Concluding........(N)	In summary......(N) To summarize...(N) Summarizing....(N)
In summing up.....(I) To sum up..........(I) Summing up.........(I)	In recapitulation....(F) To recapitulate......(F) Recapitulating......(F)

(The verbs are regular.)

For example:
"**In conclusion**, it can be seen that despite the recent publications, there have been few significant changes in our understanding of the biomechanics of swimming."
"**To summarize**, there are six dominant risk factors for the development of atherosclerosis."
"**Recapitulating**, our analysis of the performance of their graduates indicates that the top three North American Universities are Harvard, Yale and Columbia."

"**To sum up**, I predict that that it will become more difficult to gain funding for our department's research."
Here, the use of the phrasal verb lowers the level of formality and produces a phrase that can end an informal presentation or discussion.

v) **Finally** is usually used to refer to the last point in a discussion or to introduce a witticism (joke) that frequently ends many presentations. Alternatively the presentation could be ended by any of the following phrases or combination of phrases:

"Thank you for listening." (N)
"I would like to finish here, thank you for listening." (F)
"That was the last slide. Are there any questions?" (N)
"I'm afraid that we have run out of time, so we will have to stop here. Do you have any questions?" (F)

(I) = *Informal*
(N) = *Neutral*
(F) = *Formal*

UNIT 43: PRESENTATIONS II (EXERCISES)

A) Complete the following sentences with vocabulary from this unit or the previous unit.

For example: As you can ______ from the last slide, the major points are _________.
As you can **see** from the last slide, the major points are **summarized**.

1) Now I ______ like to consider the demographics of this disease. ______ at Table 1 we can see an increasing incidence.

....

2) To ______ our findings, no statistically significant data have been elicited, however several trends were noted.

....

3) _______, I would like to thank you for listening. Are there any questions?

....

4) As you can see, this table summarizes the raw morbidity and mortality data. _____ I would like to analyze the data in more detail.

....

5) ______ I ask you to look at Figure 7? This is a plot of reaction rate versus time.

....

6) So how can we synthesize this dye? If you ______ to the flow chart, then the answer becomes clear.

....

7) To ______ _____, it is clear that despite the theoretical progress that has been made in the development of nuclear fusion, we are many years from building a working reactor.

....

....

B) Do the following sentences show the correct use of phrases employed in presentations?

For example:

In conclusion, I would like to summarize the positive results of our research.............. ☑ (Correct)
To concluding, I would like to summarize the positive results of our research............. ☒ (Incorrect)

1) To sum in, the latest quality control results for this electronic component are entirely unsatisfactory.
☐ (Incorrect) ☐ (Correct)

2) Secondly, we can see that the success of this perfume can be attributed to the similarity of its chemical structure to female pheromones.
☐ (Incorrect) ☐ (Correct)

3) If we turn to this slide, we can see that the important physical properties of this alloy are listed.
☐ (Incorrect) ☐ (Correct)

4) Okay, I started this presentation by discussing the formation and behaviour of snow. Then I discussed rain in a similar fashion. Now I'm going to talk about cloud formation and behaviour.
☐ (Incorrect) ☐ (Correct)

5) Looking with table 3, it can be seen that there is a great deal of variation in the relative financial success of the biological and physical science research divisions.
☐ (Incorrect) ☐ (Correct)

6) To recapitulate, the most reliable ECG change after a myocardial infarction is ST elevation.
☐ (Incorrect) ☐ (Correct)

7) Ladies and gentlemen, would you now turn in Figure 4.
☐ (Incorrect) ☐ (Correct)

8) Now we need to address the issue of antigenic shift in this virus.
☐ (Incorrect) ☐ (Correct)

9) Finally, would you like to ask any questions?
☐ (Incorrect) ☐ (Correct)

10) To concluding, I would like to thank you for listening. Are there any questions?
☐ (Incorrect) ☐ (Correct)

UNIT 44: PRESENTATIONS III

All presentations involve the communication of information, and inevitably most also involve the communication of an opinion. The latter may include commenting on the factual content of the presentation or offering a hypothesis or theory. There are several relevant constructions that are used in scientific, technical and medical English:

A) Opinion

Think (N) Used to express an opinion.
We **think** that chemotherapy before surgery would be the best management for this group of patients.
I **think** that there may be more efficient ways of producing the Grignard Reagent.

Believe (N) Used to express a conviction or principle.
Do you **believe** that the Pythagorean theorem has been shown to be true?
I do not **believe** that the statistical analysis was appropriate for this set of data.
We **believe** that if the traffic trends continue as we have described, the maximum capacity of the roads will be exceeded in 3 years.

Feel (I) Used to express a perception. It is more informal than *believe* or *think*.
We **feel** that this interpretation may be an oversimplification.
Most doctors **feel** that the precise balance of prescribed medications in ischaemic heart disease, should be a matter of discretion for the doctor concerned.

As I see it (I) Used to express an informal opinion. This is used more often in speech than in writing.
As I see it, our methodology and conclusions are the most reliable currently available.
As we see it, the disadvantages of this new surgical technique clearly outweigh the advantages.

In my/our opinion (F) Used to express a formal opinion.
In our opinion, the interpretation of the patient longevity versus socioeconomic status data is critical in deciding the allocation of financial resources.
In my opinion, the correlation between a high intelligence quotient (IQ) and academic success is both strong and causative.

B) Hypothesis

Suggest (N) To propose, to offer for consideration. To imply.
We would like to **suggest** an alternative interpretation of this research group's results.
I **suggest** that the reason for the positive long term effect of pravastatin on individuals with ischaemic heart disease, is its ability to decrease the size of atheromatous plaques.
This information **suggests** that your theory is at best incomplete.

Hypothesize or theorize (F) To assert a hypothesis or theory.
Should they have **hypothesized** with so little evidence?
The discrepancy between the mass of the products and the energy released during the reaction caused us to **theorize** that an undiscovered subatomic particle had been created.

Postulate (F) To contend or assert a position that is assumed to be true. **Postulate** is more formal than *suggest*.
We **postulate** that the original solar system had 10 planets.
Is it logical to **postulate** that the universe has always existed?

(I) = ***Informal*** **(N)** = ***Neutral*** **(F)** = ***Formal***

UNIT 44: PRESENTATIONS III (EXERCISES)

A) Change the *italicized* words to another word or phrase with a similar meaning. For example:

Our analysis of the same data **suggests** another mechanism of action of the drug *ecstasy*.
Our analysis of the same data **implies** another mechanism of action of the drug *ecstasy*.

1) As this presentation indicates, our randomized controlled trial only looked at the common consequences of cigarette smoking. I *think* that we should extend the study to include rarer consequences.

....

2) In the 1960s Bradford Hill *asserted* criteria for disease causation in epidemiology.

....

3) The last slide describes the numerous advantages of carbon technology. Because of these advantages, I *believe* that the silicon transistors in your computer will be replaced in ten years by transistors based on carbon nanotubes.

....

4) We have reviewed all of the major articles that assess the effect of cell phone use on biological tissue. *As we see it*, the evidence for any damage to biological tissue is weak.

....

5) Recent studies have shown that the Mediterranean diet protects against depression. We *hypothesize* that it is the overall pattern of the diet rather than any one component that is important in preventing depression.

....

6) The computer's Graphics Processing Unit has much more computing power than the Central Processing Unit. I *suggest* that we will be able to double the processing power of the computer by using the Graphics Processing Unit more efficiently.

....

7) *Formal Scenario:*
It is well known that complex sugar sequences are important in biological systems. However the study of these sugars has been held back by the difficulty in synthesizing these sugars in the laboratory. *In my opinion*, the automated synthesis of such sugars is now possible.

....

B) Complete the following sentences by choosing the best answer from the offered vocabulary. For example:

I __________ (*believe/postulate/theorize*) that atomic clocks will always provide the most accurate measure of time.
I **believe** that atomic clocks will always provide the most accurate measure of time.

1) We ____________ (*theorize/in our opinion/feel*) that global warming is caused by carbon dioxide accumulation in the atmosphere but the evidence for this hypothesis is not conclusive.

....

2) Drug-receptor interactions were originally described by the "lock and key" hypothesis. This was then replaced by the "induced fit" hypothesis. Because of the evidence that I have presented, I strongly _____________ (*imply/believe/postulate*) that our theory is the most accurate description of drug-receptor interactions.

....

3) *Formal scenario:*
In the 17th century Sir Isaac Newton __________ (*postulated/suggested/thought*) and published his three laws of motion. Our research attempts to refine these postulates.

....

4) *Informal scenario:*
The theory behind using bacteria to generate electricity in a laboratory is well established. Now, ____________ (*in my opinion, as I see it, hypothesize*) there is very little preventing us from using microbial respiration to generate electricity on a large scale!

UNIT 45: GRAPHS

A) Describing the direction of change

Direction	Verb	Noun
	To increase To rise To go up (I) To climb To jump up* To shoot up* To escalate	An increase A rise ----------- A climb A jump* ----------- An escalation
	To decrease To fall To decline To go down (I) To drop To plunge* To plummet*	A decrease A fall A decline ----------- A drop A plunge* A plummet*
	To stabilize To be steady/constant To remain stable/ To be stable To level out/off	A stabilization ----------- ----------- ----------- -----------
	To fluctuate To vary To oscillate	A fluctuation A variation An oscillation

* **These are for *large* changes** (I) = Informal

For example:

The costs **will increase** in parallel with inflation.
The oscilloscope shows a **fluctuating** signal.
The size of the research grant **has remained stable** for three years.
As the stone dropped through the air its speed **went up** until it reached terminal velocity.
The population of animals in that area **shot up** to five times its original number and then **plummeted** after a predator was introduced.
The solubility of the compound **falls** as you **increase** the temperature.
After rapid initial sales of the product, sales **levelled off** at 500 units per year.

B) Prepositions used to describe changes: From, To, By, Of, At

i) **From.....to**

This construction is used to describe the change between two **known** points:
The temperature of the solution dropped **from** 40 **to** 25°C.
During the 1980s the cost of solar cells fell **from** \$15 per watt **to** \$4.5 per watt.
The heart rate increased **from** 60 **to** 120 beats per minute.

ii) **By/of**

These prepositions describe the **difference** between two positions:
The temperature of the solution dropped **by** 15°C.
During the 1980s the cost of solar cells fell **by** \$10.5 per watt.
The heart rate increased **by** 60 beats per minute.
There was a fall **in** the temperature **of** 15°C.
There was an increase **in** cost **of** \$10.5 per watt.
There was an increase **in** heart rate **of** 60 beats per minute.

Note the construction: **There was.........in.........of.........**

iii) **At**

This preposition is used to describe **a particular point**:
At 200 mph the car is not stable. **At** 70°C proteins denature.

UNIT 45: GRAPHS (EXERCISES)

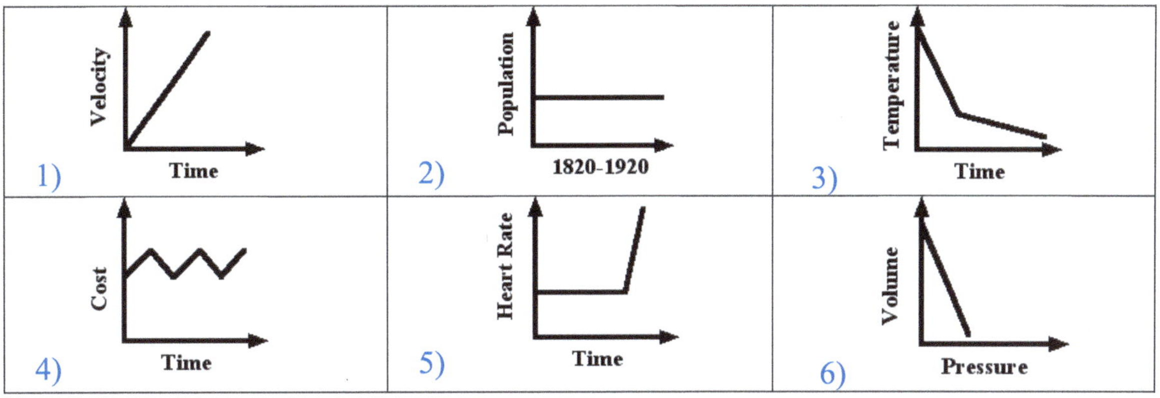

A) Using the above graphs and the vocabulary in Section A, complete the sentences below:

1) We can see from the graph that there was a very large increase in velocity with time. In fact, the velocity .... up over the period of this experiment.
2) Considering the world population between 1820 and 1920, the total number of people did not change much. The number was .... at about 1 billion.
3) The temperature change occurred in two phases. Initially the temperature .... by 70% and then .... by 20%.
4) The cost of a fluorimeter has not been constant. In recent years there has been a significant ....in prices.
5) The patient's measured heart rate was .... for six hours. Then it suddenly ...., so I suspect an infection.
6) As the pressure on the fixed mass of gas increases, the volume .... The findings are consistent with Boyle's law.

B) Rephrase the sentences below in two different ways, in a manner similar to the example:

From 1900 to 2000 the average global temperature increased from 13.2°C to 13.95°C

=>There was an increase in average global temperature of 0.75°C.

=>The global temperature increased by 0.75°C.

1) Between yesterday and today the number of colony forming units increased from 73 to 135.

....

....

2) From 2005 to now his body mass index rose from 20 to 30.

....

....

3) Over the last week the humidity has decreased from 55% to 35%.

....

....

4) The protons are accelerated by a magnetic field and they increase their energy from 10MeV to 100MeV.

....

....

5) As the ambulance passed by, the sound level dropped from 110 decibels to 30 decibels.

....

....

6) Over the next 10 years the maximum capacity of a hard disk will climb from 2 TB to 100 TB.

....

....

7) The ship took 5 minutes to decrease in speed from 40 miles per hour to 0 miles per hour.

....

....

UNIT 46: GRAPHS II

A) Slopes of graphs (gradients of graphs)

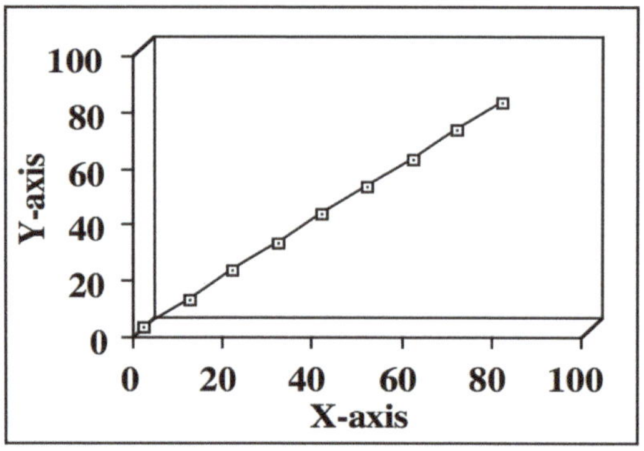

A ***verb*** + ***adverb*** can be used to describe the change in the above graph. For example:

Y climbed sharply. Y has risen dramatically. Y increases rapidly.
Y went up quickly. Y rose considerably.

Alternatively the same changes can be described using **there + to be**:

There was a sharp climb in Y.
There has been a dramatic rise in Y.
There is a rapid increase in Y.

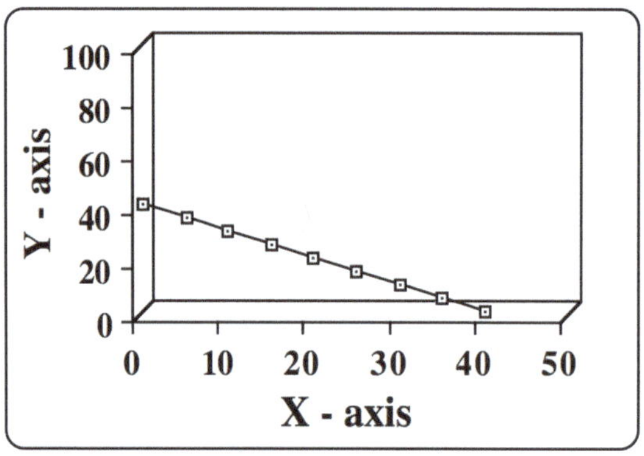

The above graph can be similarly described:

Y has declined slightly.	There has been a slight decline in Y.
Y falls gradually.	There is a gradual fall in Y.
X falls slowly.	There is a slow fall in Y.

B) A summary of adjectives and adverbs commonly used to describe changes in a graph:

Adjective	Adverb	Adjective	Adverb
constant	constantly	substantial	substantially
steady	steadily	quick	quickly
gradual	gradually	rapid	rapidly
regular	regularly	sharp	sharply
even	evenly	significant	significantly
		striking	strikingly
		great	greatly
		dramatic	dramatically

C) Small changes

To describe *small* changes, **small**, **slight** and **slightly** and **a little** may be used. For example:
There was a slight increase in camera sales this year.
Camera sales increased slightly this year.
We will need to raise the frequency a little.

UNIT 46: GRAPHS II (EXERCISES)

Rewrite the following sentences (1-23) whilst keeping the same meaning. For example:

There was a **substantial increase** in diabetes mellitus incidence this year.
Diabetes mellitus incidence **increased substantially** this year.

1) There was a dramatic increase in the rate of newspaper production.

……………………………………………………………………………………………………

2) The fluorescence of this compound will decrease gradually each year.

……………………………………………………………………………………………………

3) There has been a slight decline in the value of these cars.

……………………………………………………………………………………………………

4) The income from these shares has fallen constantly each year.

……………………………………………………………………………………………………

5) Was there a rapid increase in temperature of the solution?

……………………………………………………………………………………………………

6) The urine output of the patient rose sharply two days after the operation.

……………………………………………………………………………………………………

7) Was there a striking decrease in solubility as the temperature decreased?

……………………………………………………………………………………………………

8) There was a regular decline in pressure as the plane rose.

……………………………………………………………………………………………………

9) After sunrise, the number of photons reaching each leaf increases rapidly.

……………………………………………………………………………………………………

10) There was a dramatic decrease in EEG activity at 8pm; we believe that this was when he died.

……………………………………………………………………………………………………

11) There was a striking increase in the conductivity of the amalgam after raising the copper content.

……………………………………………………………………………………………………

12) The traffic flow fell substantially when the parallel road was closed.

……………………………………………………………………………………………………

13) The incidence of Chlamydia has been rising steadily in the under-25s.

……………………………………………………………………………………………………

14) What would happen if there were a sharp fall in the sale price of the Nikon microscope?

……………………………………………………………………………………………………

15) The ratings for this television show rose dramatically after the lead actor was changed.

……………………………………………………………………………………………………

16) The inflation rate in this country has dropped regularly every year for the last 10 years.

……………………………………………………………………………………………………

17) It is predicted that in three years time there will be a rapid increase in flash drive capacity.

……………………………………………………………………………………………………

18) The temperature of a hot liquid falls quickly initially.

……………………………………………………………………………………………………

19) The melting point of a hydrocarbon rises steadily as the carbon chain length increases.

……………………………………………………………………………………………………

20) The wind resistance does not increase sharply with increasing velocity.

……………………………………………………………………………………………………

21) There was a substantial drop in her blood pressure immediately after the operation.

……………………………………………………………………………………………………

22) The Doppler ultrasound showed that there was a significant decrease in blood flow at the carotid arteries.

……………………………………………………………………………………………………

23) Does the velocity of sound in cold air increase dramatically or decrease dramatically?

……………………………………………………………………………………………………

UNIT 47: GRAPHS III

A) The table below shows some important vocabulary:

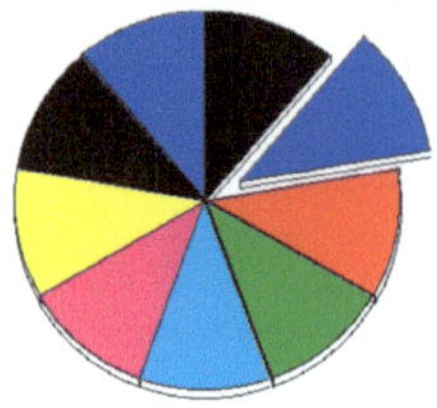

Pie Chart

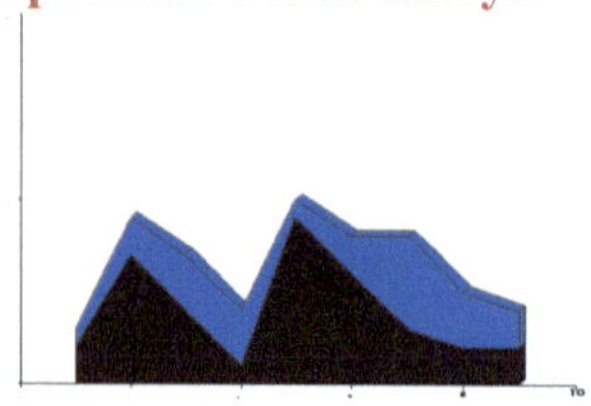

Area Graph

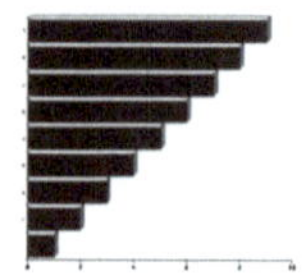

Bar Chart or Bar Graph

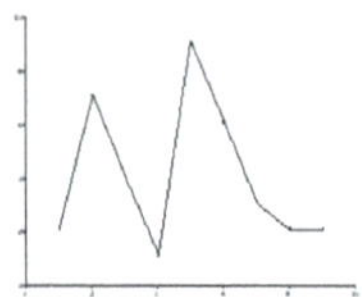

Line Graph

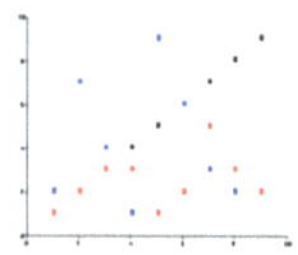

Scatter Graph

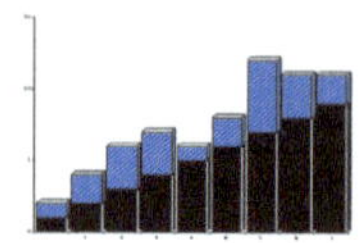

Stack Column Graph

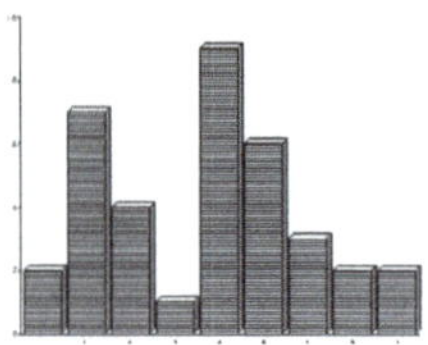

Column Graph

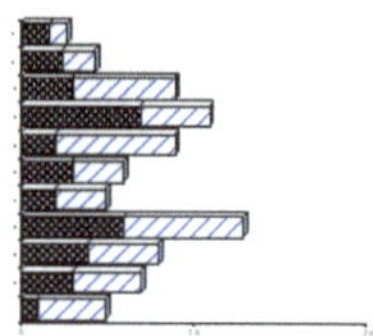

Stack Bar Graph

Polar Graph

Simple **bar charts** or **graphs** are often used for qualitative or categorical data. A **histogram** is a specialized type of bar chart that is used for continuous variables; it is used for presenting a frequency distribution.

B) Cause or Association

The following sentences contain phrases useful in technical discussions involving associations or causality:

There is a strong **association between A and B**. There is a strong correlation between A and B.
There is a **specific** association between A and B.
An association between A and B has been **consistently** demonstrated.
The **exposure to** A consistently precedes the outcome B.
Increasing the exposure to A increased the **chance** of the outcome B.
Removing the exposure A reduces the chance of the outcome B.
Similarity to results in other areas suggests that A causes B.
Based on current scientific knowledge, it is **plausible** that A would cause B.
This **analogy** is consistent with A causing B.

Is increasing age the cause of colorectal cancer or is it just a confounding factor?
Is increasing age an **association with**, or cause of, colorectal cancer?
The **data** only **shows** that height is positively correlated with weight. *(Spoken English)*
The **data** only **show** that height is positively correlated with weight. *(Written English)*

C) Statements of significance

This result is **statistically significant**. These results are statistically significant.
This result is statistically significant at the P <0.01 level.
The null hypothesis is rejected.
The results show that **the null hypothesis is rejected**.
The results indicate that the null hypothesis is rejected.

UNIT 47: GRAPHS III (EXERCISES)

Complete the sentences below by using vocabulary from the previous page, for example:

The results shown on this bar chart cannot be used to show an __________ between an individual's height and wealth.
The results shown on this bar chart cannot be used to show an **association** between an individual's height and wealth.
Your data looks impressive at first sight, but are your results actually __________ significant?
Your data looks impressive at first sight, but are your results actually **statistically** significant?
Is a Funnel plot a form of scatter _________?
Is a Funnel plot a form of scatter **graph**?

1) Obviously there is an ____________ between the popularity of a website and the number of downloads.
2) The line graph seems to show that __________ to radiation increases the risk of lung cancer.
3) Do you think that mobile phone use _________ cancer or is associated with cancer?.
4) Was a pie _______ the best way to present your data?
5) The histogram shows that the frequency of herpes infections increases with age. What if age is just a __________factor?
6) Stefan's law states a strong __________ between the temperature of a black body and the total energy radiated. The relationship is elegantly displayed on this line ___________ of temperature against radiation.
7) If you compare last year's pie __________ with this year's ________ _______, there is a clear drop in the proportion of the company's profits spent on this division.
8) Hooke's law states that the force applied is directly proportional to the extension. Are the ____________ presented on the graph consistent with this?
9) Would it be better to present the data on an area graph or a stack _______ _______?
10) If you plotted a _________ of microchip processor speed against the year, you would see a line going up rapidly.
11) Can we use the line ___________ of inflation rate against time, to predict next year's inflation rate?
12) Our results are _________ ________. In fact there is a 99.9% chance that they did not occur by chance.
13) The data do not support rejection of the ____________ hypothesis. However a larger study is needed.
14) The bar ___________ shows that the median group for male height is 175 cm – 180 cm.
15) If your results satisfy the Bradford-Hill criteria then __________ is more likely than just an association.
16) He is one of the few remaining people that believe smoking does not ___________ cancer.
17) Your graph shows there is a ___________ between individual affluence and individual longevity.
18) The slope of your graph of distance against time represents speed. Is there a statistically __________ change in the speed?
19) What type of ___________ would be best for demonstrating Michaelis-Menten enzyme kinetics?

Common Scientific, Technical and Medical Abbreviations

The abbreviations on the following pages represent a compilation of terms that a student of English is likely to see and use in their work. The student should be aware that each workplace may also have its own abbreviations that are beyond the scope of this text.

Common Scientific, Technical and Medical Abbreviations

A

a_0 Bohr radius
a absorption coefficient; attenuation coefficient; acceleration, angular acceleration; fine structure constant; helium nucleus
A Helmholtz free energy; radioactive activity; angle of prism; area; mass number; nucleon number; mean distance from the earth to the sun; adenine; ampere; absolute temperature; ankle reflex
Å angstrom
A_H Hall coefficient
A_r relative atomic mass of an element
A- general symbol for an anion
AAA Abdominal aortic aneurysm
A & AEE Aeroplane and Armament Experimental Establishment
AAL anterior axillary line
AAS atomic absorption spectroscopy
AASH adrenal androgen stimulating hormone
ab as before
AB apex beat
Abe abequose
ABA abscisic acid
ABC airways, breathing, circulation
abd / abdo. abdomen
ABG arterial blood gas
ABO blood classification system (A, AB, B, O)
ABP androgen binding protein
abs absolute
ABS Acrilonitrile-Butadiene-Styrene
ABS brake anti-lock brake
a.c. alternating current
Ac actinium; transformation temperature; before meals, acetyl
ACE angiotensin converting enzyme
ACL anterior clavicular line
AcOH acetic acid
ACM Association for Computing Machinery
ACP acyl carrier protein
ACR approach control radar
ACS active control system; attitude control system; air conditioning system
ACT active control technology; activated complex theory
ACTH adrenocorticotrophic hormone
ACV air cushion vehicle
A/D analogue-to-digital converter
ADC analogue-digital converter
ADD airstream direction detector
Ade Adenine
ADF automatic direction finding
A-DNA A-type deoxyribonucleic acid
ADH antidiuretic hormone
ad lib at will, as desired
adm admission
ADP adenosine diphosphate; automatic data processing
ads adsorption
ae eon (10^9 years)
Ae transformation temperature at equilibrium
AE air entry; Accident and Emergency
AERO Air Education and Recreation Organisation (U.K.)
AES atomic emission spectroscopy; Auger electron spectroscopy
AF\a.f. audio frequency, atrial fibrillation
AFC automatic frequency control
AFCS automatic flight control system
AFM atomic force microscopy
AFP alphafetoprotein
Ag silver
AGC automatic gain control
AGR advanced gas cooled (nuclear) reactor
AHG antihemophilic globulin
AI artificial insemination; artificial intelligence; aortic incompetence
AIAA American Institute of Aeronautics and Astronautics
AIBN azoisobutyronitrile
AID artificial insemination by donor

AIDS acquired immune deficiency syndrome
AIH artificial insemination by husband
AIM atoms in molecules
AJ ankle jerk/reflex
Al aluminium; Alfen number
Ala alanine
ALARA as low as reasonably achievable
alc alcohol
aliph aliphatic
alk alkaline
alk. phos. alkaline phosphatase
All allose
Alt altrose
ALT alanine aminotransferase / transaminase
ALU arithmetic logic unit
am amorphous solid
Am americium; amyl
AM amplitude modulation
amp ampere; amplifier
AMP adenosine monophosphate
AN antenatal
ANF atrial natriuretic factor
ANP atrial natrurietic peptide
ANS autonomic nervous system
ANSI American National Standards Institute
anti- opposite face
AoR Aortic valve replacement
AP alkaline phosphatase, antero-posterior
APH antepartum haemorrhage
API American Petroleum Institute
APTT, aPTT Activated Partial Thromboplastin Time
APU auxiliary power unit
APUD amine precursor uptake and decarboxylation cells
aq aqueous
Ar- aryl
Ar argon
AR aspect ratio; analytical reagent
Ara arabinose
ARAS ascending reticular activating system
ARB Air Registration Board
ARC Agricultural Research Council
Arg arginine
ARM anti-radiation missile, artificial rupture of membranes
ARPA Advanced Research Projects Agency
as asymmetrical
As arsenic
AS abdominal system; aortic stenosis
ASA American Standards Association
ASAP as soon as possible
ASCII American Standard Code for Information Interchange
ASD atrial septal defect
ASE aromatic stabilization energy
Asn asparagine
ASO antistreptolysin O
Asp aspartate, aspartic acid
asym asymmetrical
at atomization
At astatine
ATB aeration test burner
ATC air traffic controller
ATCRBS air traffic control radar beacon system
atm atmosphere
ATP adenosine triphosphate
ATPase adenosine triphosphatase
ATR attenuated total internal reflection
ATS antitetanic serum, antitetanus serum
at.wt. atomic weight
au astronomical unit/mean distance from the Earth to the Sun
Au gold
av average
A-V arteriovenous concentration difference of any given substance
AV atrioventricular
AVC automatic volume control
avdp avoirdupois
AVF augmented voltage foot
avgas aviation gas
AVL augmented voltage left arm
AVM arteriovenous malformation
avpin aviation isopropyl nitrate
avpol aviation petrol, oil and lubricant.
AVR augmented voltage right arm
avtag aviation wide-cut turbine fuel.
avtur aviation turbine fuel

AUT Association of University Teachers
a&w alive and well
AWACS airborne warning and control system
AXR abdominal x-ray
AZT azidothymidine

B

b breadth
b angular momentum; barn
ß phase constant; ratio of velocity to velocity of light
B boron; best; magnetic flux density; second viral coefficient; susceptance; biceps reflex
Ba barium
BA Bachelor of Arts; British Association screw thread; British Airways
BAL British anti-lewisite
B-Alkyl-9-BBN B-Alkyl-9-borabicyclo[3.3.1]nonane
BALPA British Airline Pilots Association
BALUN Balance to unbalance transformer
bar bar (pressure unit)
BASF Badische Anilin and Soda Fabrik
BASIC beginner's all purpose symbolic instruction code
BB best best, bed bath, blanket bath
BBB bundle branch block
bbl barrel
9-BBN borabicyclo[3.3.1]nonane
BC bayonet cap
BCC body centered cubic; basal cell carcinoma
BCD binary coded decimal
BCF bromochloro difluoromethane
BCG bacille Calmette Guerin
BCS British Computer Society
BCS theory Bardeen, Cooper and Schrieffer theory of superconductivity
bd twice a day
BD Blu-ray Disc
BDE bond dissociation energy
B-DNA B-type deoxyribonucleic acid
BDV breakdown voltage
Be beryllium
Bé Baumé
BE Bachelor of Engineering
BER basic electrical rhythm
BET Brauner, Emmett and Teller adsorption theory
BeV billion electronvolts
B-field magnetic field
BFO beat frequency oscillator
BGA British Gliding Association
BHA bottom-hole assembly
BHN Brinell Hardness Number
b.h.p. brake horsepower
Bi bismuth; biot
Bid/bid twice a day
BID brought in dead
BIOS basic input output system
Bit **bi**nary dig**it**
BITNET because its time network
Bk berkelium
BLEU blind landing experimental unit
Bm bowel movement
BM benchmark
BMEP brake mean effective pressure
BMI Body mass index
BMR basal metabolic rate
BN bond number
BNF Bachus-Naur form; British National Formulary
BNO bowels not opened
BNS nuclear backscattering spectroscopy
BO bowels opened; bond order
BOD biochemical oxygen demand
BOP blow out preventer
BOR bowels open regularly
b.p./bp boiling point
BP British Petroleum; blood pressure; between perpendiculars
BPC British Pharmaceutical Codex
BPd diastolic blood pressure
BPD bi-parietal diameter
BPs systolic blood pressure
Bq becquerel
Br bromine
BRE bond resonance energy
BRM binary rate multiplier
BRP bathroom privileges
BS British Standard, breath sounds, bowel sounds

BSc. Bachelor of Science
BSA Bovine serum albumin
BSE Bovine spongiform encephalopathy
BSP sulfobromophthalein
BSSE basis set superposition error
B.th.u./Btu British thermal unit
BTU board of trade unit
bu bushel
Bu butyl
BUN blood urea nitrogen
BWG Birmingham wire gauge
BWR Boiling Water (nuclear) Reactor
BWt Birth weight
Bz benzoyl

C

c concentration; speed of light; specific heat capacity; combustion reaction; velocity
c. circa (L) = about
C carbon; centigrade/celsius; component; coulomb; heat capacity; cytosine; complement; number concentration
^{o}C degree Celsius
c_0 speed of light in a vacuum
c_p specific heat capacity at constant pressure
c_v specific heat capacity at constant volume
ca approximately
Ca calcium; cancer
CA cancer; carcinoma
CAA civil aviation authority
CAB civil aeronautics board; cellulose acetate butyrate
CABG coronary artery bypass graft
CAD computer aided design
CAE computer aided engineering
CAI computer assisted/aided instruction
cal calorie
Cal kilocalorie
CAL computer assisted learning
calc calculated
CAM computer aided machining
cAMP cyclic adenosine 3',5'-monophosphate
Cap capsule
CARS coherent anti-Stokes Raman spectroscopy
CAS collision avoidance system
CAS RN Chemical Abstracts Service Registry Number
CAT computer assisted/aided testing; computerized axial tomography/computer assisted tomography; clear-air turbulence
CATV community antenna television
CB citizen's band radio
CBF cerebral blood flow
CBG corticosteroid binding globulin/transcortin
CBS complete basis set
CBT computer based training, cognitive behavioural therapy
cc cubic centimeter, with meals
CC chief complaint
CCD charge coupled device
CCF congestive cardiac failure
CC filter colour correction filter/colour conversion filter
CCIR Comité Consultatif International des Radiocommunications
CCITT Consultative Committee International Telegraph and Telephone
CCK cholecystokinin
CCP cubic close packing
CCR cassette camera recorder
CCTV closed circuit television
CCV control-configured vehicle
cd candela; current density
Cd cadmium
CD compact disk; circular dichroism
cDNA complementary deoxyribonucleic acid
CDP cytidine diphosphate
Ce cerium
CEPA couplet electron pair approximation
CET Central European Time
cf. compare (from L. conferre)
Cf californium
CFA complete Freund's adjuvant
CFF critical fusion frequency
cfm cubic feet per minute
CFS chronic fatigue syndrome

CFT complement fixation test
CGA colour/graphics adapter
cGMP cyclic guanosine 3',5'-monophosphate
CGP chorionic growth hormone-prolactin
CGRP calcitonin-gene-related peptide
CGS centimetre-gram-second
CHF coupled Hartree-Fock (method)
chip microchip
Chr. CF chronic cardiac failure
Ci curie
CI configuration interaction; chemical ionization
CIDEP chemically induced dynamic electron polarization
CIDNP chemically induced dynamic nuclear polarization
CIE Commission Internationale d'Éclairage
CI engine compression-ignition engine
CKFF Cotton-Kraihanzel force field
Cl chlorine
CL cathode luminescence
CLIP corticotropinlike intermediate lobe peptide
cm centimeter; centre of mass
Cm curium
cmc critic micelle concentration
CMI computer managed instruction
CMO canonical molecular orbital
CMOS complementary metal-oxide-semiconductor
CMP cytidine monophosphate
CMR carbon-13 nuclear magnetic resonance
$CMRO_2$ cerebral metabolic rate for oxygen
CMV cytomegalovirus
CN coordination number
CN^- cyanide
CNAA Council for National Academic Awards
CND Campaign for Nuclear Disarmament
CNDO complete neglect of differential overlap
CNES Centre National d'Etudes Spatiales
CNS central nervous system
c/o complains of
Co cobalt; Cowling number
CoA coenzyme A
COAD chronic obstructive airways disease
COBOL common business orientated language
COD chemical oxygen demand
CODAN carrier operated device anti-noise
C of A certificate of airworthiness
COHb carboxyhemoglobin
C of M certificate of maintenance
COM computer output on microfilm /microfiche
COMT catechol-o-methyltransferase
conc concentrated, concentration
const constant
COP change of plaster
COPD chronic obstructive pulmonary disease
CoQ coenzyme Q (ubiquinone)
cos cosine
cosec cosecant
cosh hyperbolic cosine
COSY correlation spectroscopy (in NMR)
cot cotangent
coth hyperbolic cotangent
c.p./cp candlepower; centre of pressure
cP centipoise
Cp cyclopentadienyl
CP chemically pure; change point
CPA coherent potential approximation
cpd contact potential difference
Cpd. compound
CP filter
CPL commercial pilot's licence
CP/M control program monitor
CPM critical path monitor
CPR cardiopulmonary resuscitation
cps cycles per second
CPT charge conjugation-space inversion-time reversal (theorem)
CPU central processing unit
cr crystalline phase
Cr chromium; creatinine
CR conditioned reflex

creps crepitations
CRF corticotrophin-releasing hormone
CRH corticotrophin-releasing hormone
CRI colour reversal intermediate
CRO cathode ray oscilloscope
CrP creatinine phosphate
CRT cathode ray tube
CRU constitutional repeating unit
cryst crystalline phase
c/s cycles per second
Cs caesium
CS conditioned stimulus
csa cross sectional area
csc cosecant
CSE Certificate of Secondary Education
CSF cerebrospinal fluid; colony stimulating factor; Canadian standard freeness
CS gas orthochlorobenzylidene malononitrile
CSI compact source iodide
CSO colour separation overlay
CSSD Central Sterile Supply Depot
CSU catheter specimen of urine
ct. cent; carat
CT cerebral tumour; coronary thrombus; computerized tomography; charge transfer
CTC City Technology College
CTEM conventional transmission electron microscopy
CTP cytidine triphosphate
CTR controlled thermonuclear reactor/reaction
cu cubic
Cu copper
CUB cubic
cusec cubic foot per second
cv cultivar
CV closing volume; cardiovascular; cyclic voltammetry
CVA cerebrovascular accident; cardiovascular accident
CVD chemical vapor deposition
CVR cerebral vascular resistance
CVS chorionic villus sampling, cardiovascular system, cerebrovascular system
CVT constant variable transmission
cw continuous wave
cwt hundredweight
Cx cervix; cervical spine
CXR chest x-ray
Cy cyclohexyl
cyl cylinder
Cys cyteine
CyS half-cystine
cyt cytochrome
CZI crystalline zinc insulin

D

d density; diameter; gratings spacing or slit separation; thickness; day; atomic orbital "diffuse"; dextrorotatory; deuteron; distance
D dioptre; Deuterium; Dimension; electric displacement; electric flux density; angle of deviation; debye unit; diffusion coefficient
2-4-D 2,4-Dichlorophenoxyacetic acid
da deca-
Da Dalton
DA donor-acceptor (complex)
D/A or D-A digital to analogue
DALR dry adiabatic lapse rate
D and K damaged and kept
DAP distributed array processor
DARPA defense advanced research projects agency
DAT digital audiotape
dB decibel
DBMS data base management system
DBS direct broadcasting satellite/direct broadcasting by satellite
d/c discharge, discontinue
d.c. direct current
D&C dilatation and curettage
D&V diarrhoea and vomiting
DCF deal-cased frame
DD dangerous drugs
DDAVP 1-deamino-8-D-arginine vasopressin
DDD derivative of DDT
DDL data description languages
DDQ 2,3-dichloro-5,6-dicyanoquinone
DDT 1,1,1-trichloro-2,2-bis-(p-chlorophenyl)ethane

DE delocalization energy
DEA/DHEA dehydroepiandrosterone
dec decomposes
deca- ten
deci- a tenth
decub. lying down
deg. degree
demi- half
Dept. department
DES Department of Education and Science; data encryption standard
det determinant
dev deviation
DFP diisopropyl fluorophosphate
DFS disk filing system
DFT density functional theory
DFVLR Deutsche Forschung und Versuchanstalt für Luft und Raumfahrt
D/H drug history
DHP 2,3-dihydro-4H-pyran
DHT dihydrotesterone
di- two/twice/double
dia. diameter
DIA diamond structure
diam diameter
dil dilute, dilution
DIM diatomics in molecules
DIN Deutsche Institut fr Normung
DIP dual in-line package; dorsal interphalangeal joint
DIT diiodotyrosine
dl deciliter
DLE discoid lupus erythematosus
dm decimetre
DM disconnecting manhole
DMDT methoxychlor
DME distance-measuring unit
DMF dimethylformamide
DMSO dimethylsulphoxide
D/N ratio ratio of dextrose to nitrogen in the urine
DNA deoxyribonuleic acid, did not attend (or arrive)
DNAse deoxyribonuclease
DNMR dynamic nuclear magnetic resonance
DOA dead on arrival
DOB date of birth
DOS density of states
doz dozen
dp degree of polymerization
DPhil doctorate (doctor of philosophy)
dpl displacement
DPL dipalmitoyl lecithin
dpm disintegrations per minute
DPN diphosphopyridine nucleotide
DPNH reduced diphosphopyridine nucleotide
DPPC dipalmitoyl phosphatidyl choline
dps disintegrations per second
dr. dram
dRib 2-deoxyribose
DRIFT diffuse reflectance infrared Fourier transform
DRS diffuse reflectance spectroscopy
DS disseminated sclerosis
DSC differential scanning calorimetry
D.Sc. Doctor of Science
DTA differential thermal analysis
DTL diode transistor logic
DTP desktop publishing
DTR diffusion-transfer reversal
DTs delirium tremens
DU duodenal ulcer
D&V diarrhoea and vomiting
DVD Digital Versatile Disc, Digital Video Disc
DVT deep venous thrombosis
Dx diagnosis
DXT deep x-ray treatment
Dys dysprosium
dyn. see dyne
dyne a unit of force (causes an increase in velocity of a gram object of one centimetre per second, every second)

E

e electronic charge; linear strain
e the base of natural logarithms
E electrolytes; Electric field strength, electromotive force; energy; modulus of elasticity; entgegen (trans configuration)
E_h Hartree energy
EA electron affinity
EACA epsilon-aminocaproic acid

EAM external auditory meatus
EAN European Academic Network
EARN European Academic Research Network
EBL estimated blood loss
EBM Electron Beam Machining
EBNA Epstein-Barr virus Nuclear Antigen
EBR electron beam recording
EBU European Broadcasting Union
EBV Ebstein Barr Virus
EBW electron beam welding
ECAC European Civil Aviation Conference
E1cb elimination unimolecular of the conjugate base
ECC earth continuity conductor
ECF extracellular fluid
ECFA eosinophil chemotactic factor of anaphylaxis
eCG equine chorionic gonadotropin
ECG electrocardiogram
ECM electrochemical machining; electronic counter measures, extracellular matrix
ECoG electrocorticogram
EcoRI EcoRI restriction endonuclease
ECP effective core potential
ECR electron cyclotron resonance
ECT electroconvulsive therapy
ecto- outside
ED electron diffraction, effective dose
EDC expected date of confinement
EDD expected date of delivery
EDM early diastolic murmur
EDP electronic data processing
EDRF endothelium derived relaxation factor
EDS energy dispersive X-ray spectroscopy
EDTA ethylenediaminetetraacetic acid
EEG electroencephalogram
EELS electron energy loss spectroscopy
EFFF energy factored force field
E-field Electric field
EFP electronic field production
EFT electronic funds transfer
EGA enhanced graphics adapter
EGF epidermal growth factor
EGT exhaust gas temperature
EHF extremely high frequency
EHMO extended Hückel molecular orbital
ehp equivalent break horsepower
EHT extended Hückel theory
EI exposure index
EIA Electronic Industries Association
EJP excitatory junction potential
EKG electrocardiogram
ELINT electronic intelligence
ELISA enzyme-linked immunosorbent assay
ELV expendable launch vehicle
email electronic mail
EMF/e.m.f. electromotive force
EMG Electromyography, Electromyograph
EMI electromagnetic interference
emu electromagnetic unit system
EMU early morning specimen of urine
en ethylenediamine
ENDOR electron-nuclear double resonance
ENG electronic news gathering; head engaged
ENT ear, nose and throat
EOS equation of state
erg unit of energy (the work done in moving a force of 1 dyne one centimetre along the line of action).
EP endogenous pyrogen
EPMA electron probe microanalysis
EPR electron paramagnetic resonance
EPROM erasable programmable read-only memory
EPSP excitatory postsynaptic potential
eq equivalent; equation
eqn equation
eqQ quadrupole coupling constant
Er erbium
ER endoplasmic reticulum, emergency room
ERC ever-ready case
erf error function
erg erg
ERG electroretinogram
ERPF effective renal plasma flow
ERV expiratory reserve volume

Es einsteinium
ESA European Space Agency
ESCA electron spectroscopy for chemical analysis
esd estimated standard deviation
ESD electron stimulated desorption
ESN educationally subnormal
ESPRIT European Strategic Programme for Research in Information Technology
ESR electron spin resonance, erythrocyte sedimentation rate
est estimate, estimated
e.s.u. electrostatic unit
Et- ethyl-
ET electron transfer; ephemeris time
ETA estimated time of arrival
ETD estimated time of delivery
ETP electron transport particle
ETT exercise tolerance test
eu entropy unit
Eu europium; Euler number
EUA examination under anaesthesia
EUMETSAT European Meteorological Satellite Organisation
EURECA European Retrievable Carrier
EUTELSAT European Telecommunications Satellite Organisation
eV electronvolt
EV exposure value
EVA extra-vehicular activity
EW electronic wave
EWG electron withdrawing group
EWT elsewhere taken
ex- out
ex. & ct. excavate and cart away
EXAFS extended x-ray absorption find structure
EXELFS extended energy loss fine structure
exo- external
exp exponential function
expt experimental
ext. exterior/external
extra- outside

F

f formation reaction
f focal length; frequency; femto- ; atomic orbital "fundamental"; activity coefficient; aperture ratio; focal length; force constant; fugacity
F farenheit; Faraday/Farad; fluorine; degrees of freedom; force; Helmholtz free energy; activity coefficient; power of a lens; filial generation; angular momentum
Fm magnetomotive force
FAA Federal Aviation Administration
FACS fluorescence activated cell sorter
FAD flavin adenine dinucleotide
$FADH_2$ reduced flavin adenine dinucleotide
Fahr. Fahrenheit
FAI fresh-air inlet
Fb foreign body, finger breadth
FBC full blood count
FBG fasting blood glucose
FBR fast breeder reactor
FC fair cutting
FCC face centered cubic
fd framed
FD forceps delivery
FDA Food and Drugs Administration
FDD floppy disk drive
FDP fibrinogen degradation product
FDS Fermi-Dirac-Sommerfield
Fe iron
FEL free electron laser
FEM field emission microscopy
FEMO free electron molecular orbital
femto- 10^{-15}
FEP fluorinated ethene propene
FET field effect transistor
FeV1 forced expiratory volume in 1 second
FFA free fatty acid
FFT fast fourier transform
FGF fibroblast growth factor
FH Family history
FHH foetal heart heard
FHNH foetal heart not heard
FHP friction horsepower
fid free induction decay
fig figure
FIM field ion microscopy
FIR flight information region

FITC fluorescein isothiocyanate
fl fluid, femtolitre
FLIR forward-looking infra red
FLOPS floating point operations per second
fl oz fluid ounce
fm fathom
Fm fermium
FM frequency modulation
FMEA failure modes and effects analysis
fMET formylmethionine
FMFF fetal movement first felt
FMN flavin mononucleotide
FMNH2 reduced flavin mononucleotide
FMS flexible manufacturing system
Fo Fourier number
FOB faecal occult blood
FORTRAN formula translation
FP/f.p. freezing point
FP forward perpendicular
FPS Fellow of the Pharmaceutical Society of Great Britain; feet,pounds and seconds
fps feet per second; foot-pound-second
Fr francium; franklin
FRC functional residual capacity
FRH, FSH-RH, FRF follicle stimulating hormone releasing hormone
FROM full range of movement
FRS Fellow of the Royal Society
Fru fructose
FSD\f.s.d. full scale deflection
FSGO floating spherical Gaussian orbital
FSH follicle-stimulating hormone
FSK frequency-shift keying
ft foot/feet
FT Fourier transform
FTD formal thought disorder
FTIR Fourier transform infrared spectroscopy
FTND full term normal delivery
FU Follow up
Fuc fucose
FUO fever of unknown origin
fus fusion (melting)
fx fracture

G

g gram(me); gravity; osmotic coefficient; load factor; gas; acceleration due to gravity; degeneracy; statistical weight, Landé g-factor
G gravitational constant; gauss; giga-; Gibbs function/free energy; conductance shear modulus; guanine; glucose; gulley; G-value; thermodynamic potential, gravidity
Ga gallium
GA general anaesthetic
GABA gamma aminobutyric acid
GAD glutamate decarboxylase
gal gallon
Gal gal, galileo, galactose
GAPAN guild of air pilots and air navigators
GB gallbladder
GBG gonadal steroid binding hormone
GC general condition, gas chromatography
GCA ground controlled approach
GCE General Certificate of Education
GCFT gonococcal complement fixation test
GCI ground controlled interception
GC-MS gas chromatography-mass spectrometry
GCS Glasgow coma scale
GCSE General Certificate of Secondary Education
G-CSF macrophage colony stimulating factor
Gd gadolinium
GDMS glow discharge mass spectroscopy
GDP guanosine diphosphate
Ge germanium
gem geminal
geo- earth
GeV gigaelectronvolt
GFR glomerular filtration rate
GH growth hormone
GIAO gauge invariant atomic orbital
GIP glazed imitation parchment
GIS gastro-intestinal system
GIT gastrointestinal tract

gl glacial
glc glucose
GLC gas-liquid chromatography
Gln glutamine
Glu glutamate
Gly glycine
gm gram
GM-CSF granulocyte macrophage colony stimulating factor
GMP guanosine monophosphate
GMT Greenwich Mean Time
GnRH gonadotropin-releasing hormone
GP General Practitioner
GPI general paralysis of the insane
gpm gallons per minute
gps gallons per second
GPT glutamic pyruvic transaminase
gr grain
Gr Grashof number
GRH/GRF growth hormone releasing hormone
GR-S government rubber-styrene
GSE ground support equipment
GSH reduced glutathione
G6PD(H) glucose 6-phosphate dehydrogenase
GSSG oxidized glutathione
GSR galvanic skin response
GTN glyceryl trinitrate
GTO Gaussian type atomic orbital
GTP guanosine triphosphate
GTPase guanosine triphosphatase
gtt drop-liquid measurement
GTT glucose tolerance test
GU gastric ulcer
Gua guanine
GUM genitourinary medicine
GUS genito-urinary system
GUT grand unified theory
GVB generalized valence bond
GVH graft-versus-host reaction
GWS Glashow-Weinberg-Salam
Gy gray, gigayear
Gyn gynaecology

H

h Planck's constant; height; hour; Planck's constant, helion
H hydrogen; Enthalpy; henry; Earth's horizontal magnetic field; histidine, enthalpy
ha hectare
Ha hahnium; Hartmann number
HA general symbol for an acid
HAM hydrogen atoms in molecules
hav haversine
Hb haemoglobin
HBE His bundle electrogram
HbO_2 oxyhaemoglobin
HCC, 25-HCC 25-hydroxycholecalciferol
hCG human chorionic gonadotrophin
HCl hydrochloric acid
HCP hexagonal close packed
hCS human chorionic somatomammotropin
Hct haematocrit
HDD hard disk drive; head down display
HD DVD High-Definition/Density DVD
HDL high density lipoprotein
HDMI High-Definition Multimedia Interface
HDTV high definition television
He helium
HEIS high energy ion scattering
HEM hybrid electromagnetic wave
HEP high energy physics
HETP height equivalent to a theoretical plate
hex-/hexa- six
HEX hexagonal
hf. half
Hf hafnium
HF high frequency
hfs hyperfine structure
hg hectogram
Hg mercury
hGH human growth hormone
HGPRT hypoxanthine - guanine phosphoribosyl transferase
HH hiatus hernia
HHV-8 human herpes virus 8
HI height of instrument; head injury
HI arc high intensity arc
HIOMT hydroxyindole-O-methyltransferase

His histidine
HIV human immunodeficiency virus
hl hectolitre
HLA histocompatibility antigen
HLB hydrophilic-lipophilic balance
hm hectometer
HMO Hückel molecular orbital
HMPT hexamethylphosphorotriamide
HNPU has not passed urine
HND Higher National Diploma
HNW head, nut and washer
H_0 Hubble constant
Ho holmium
HOMO highest occupied molecular orbital
HOTOL horizontal take-off and landing
hp horsepower
H & P history and physical
HPC History of presenting complaint
hph horsepower hour
hPL human placental lactogen
HPLC high-performance liquid chromatography
HPV human papilloma virus
hr hour
HR heart rate
HREELS high resolution electron energy loss spectroscopy
HREM high resolution electron microscopy
HRP horseradish peroxidase
hs hour of sleep (bedtime)
HS heart sounds
HSAB hard-soft acid base (theory)
HS-CoA reduced coenzyme A
HSE homodesmotic stabilization energy
H substance histamine-like capillary vasodilator
HT/ht height
5-HT serotonin
HTR high-temperature reaction
HTU height of a transfer unit
HUD head(s)-up display
HVA homovanillic acid
HVS high vaginal swab
Hx history (of complaint)
Hyl hydroxylysine
Hyp 4-hydroxyproline
Hz hertz

I

i van't Hoff's factor; square root of minus one
I iodine; electric current; ionic strength; luminous intensity; moment of inertia; sound density; isoleucine; nuclear spin angular momentum; radiant intensity
IA international angström
IAA indole-3-acetic acid
IACS International Annealed Copper Standard
IAEA International Atomic Energy Commission
IAR instruction address register
IAS indicated air speed
IAT international atomic time
IATA International Air Transport Association
IBA indole-3-butyric acid
IBM International Business Machines
IBS irritable bowel syndrome
IC inspiratory capacity
I.C. integrated circuit
I.C.A.O. International Civil Aviation Organisation
ICF intracellular fluid
I.Chem.E. Institution of Chemical Engineers
I.C.I. Imperial Chemical Industries
ICP inductively coupled plasma
ICR ion cyclotron resonance
ICRH ion cyclotron resonance heating
ICS intercostal space
ICSH interstitial cell stimulating hormone
ICU intensive care unit
ICW interrupted continuous waves
id ideal (solution)
ID infectious disease; inside diameter
IDDM insulin dependent diabetes mellitus
IDL intermediate density lipoprotein
Ido idose
IE ionization energy
IEA International Energy Agency
IEE Institution of Electrical Engineers
iep isoelectric point

IEPA independent electron pair approximation
IF intermediate frequency
iff if and only if
IFIP International Federation for Information Processing
IFN interferon
IFR instrument flight rules
IFRB international frequency registration board
IGF-I, IGF-II Insulin-like growth factors I and II
Ig immunoglobulin
IgG immunoglobulin G
IGLO individual gauge for localized
IGT in growing toenail
IGV inlet guide vanes
IHC immunohistochemistry
i.h.p. indicated horsepower
IKBS intelligent knowledge based systems
IJP inhibitory junction potential
IL interleukin
Ile/Ileu isoleucine
ILS instrument landing system
im/IM intramuscular
IMB intermenstrual bleeding
imm immersion
IMMARSAT International Maritime Satellite Organisation
IMC instrument meteoreological conditions; integrated memory controller
I.Mech.E. Institution of Mechanical Engineers
IMEP indicated mean effective pressure
I.Min.E. Institution of Mining Engineers
IMP iodoamphetamine
IMPATT diode impact avalanche transit-time diode
I.Mun.E. Institution of Municipal Engineers
In indium; inulin
INDO intermediate neglect of differential overlap
infus^{n} infusion
INMARSAT International Maritime Satellite Organisation
INR International Normalized Ratio
INS inelastic neutron scattering, ion neutralization
Inst.P. Institute of Physics
int internal
INTELSAT InternationalTelecommunications Satellite Organisations
I & O intake and output
I/O input/output
IOFB intra-ocular foreign body
IOP input/output processor
IP in-patient; interphalangeal; ionization potential
IP_3 inositol triphosphate
IPA isopentenyl adenine
IPN interpenetrating polymer network
IPR isotopic perturbation of resonance
IPSP inhibitory postsynaptic potential
IPT Institute of Petroleum Technologists
IPTS International Practical Temperature Scale
I.Q. intelligence quotient
Ir iridium
IR/i.r. infrared
IR information retrieval
IRAS reflection-absorption infrared spectroscopy
IRC intrinsic reaction coordinate
IRCM infrared countermeasures
IRM innate releasing mechanism
IRS infrared spectroscopy
IRV inspiratory reserve volume
ISA International Standard Atmosphere
ISAS Institute of Space and Aeronautical Science of the University of Tokyo
isc intersystem crossing
ISE isodesmic stabilization energy
ISQ condition unchanged
ISRO Indian Space Research Organisation
ISS ion scattering spectroscopy
IT information technology
ITM inch trim moment
ITP inosine triphosphate
ITS International Temperature Scale (1990)
ITU International Telecommunication Union

I.U./i.u. international unit
IUD intrauterine device
IUPAC International Union of Pure and Applied Chemistry
iv/IV intravenous
IVC inferior vena cava
IVF in vitro fertilization
IVP intravenous pyelogram
IVU intravenous urogram
IW isotopic weight
Ix investigations
IX ion exchange

J

j angular momentum
J joule; electric current density; magnetic polarisation; moment of inertia; sound intensity; Massieu function
JANET joint academic network
JATO jet-assisted take-off
JCL job control language
JET Joint European Torus
JFET junction field-effect transistor
JFIF JPEG File Interchange Format
JND just noticeable difference
JPEG Joint Photographic Experts Group
JVP Jugular venous pressure

K

K electrolytic conductivity; compressibility; magnetic susceptibility
k knot; kilo- ; thermal conductivity; radius of gyration; Boltzman constant; absorption index; rate constant; wave vector
K potassium; Kelvin; 1000; equilibrium constant; bulk constant; bulk modulus; lysine; kilodalton; solubility constant; kinetic energy; absorption coefficient
K_E exchangeable body potassium
KB/kb kilobyte
kc kilocycle
kcal kilocalorie
kCi kiloCurie
kc/s kilocycles per second
KE kinetic energy
keV kilo-electronvolt
K_f glomerular ultrafiltration coefficient
kg kilogram
kgf kilogram force
kHz kilohertz
kJ kilojoule
kl kilolitre
km kilometre
kn knot
Kn Knudsen number
Kr krypton
kPa kilopascal
k.p.h./kph kilometres per hour
K_{sp} solubility product
kt karat
KUB kidney, ureter and bladder
kV kilovolt
kVA kilovolt-ampere
kVAr kilovar
kV_p kiloVolts, peak
kW kilowatt
kWh kilowatt hour
KWAC keyword and context
kWh kilowatt-hour

L

l litre; length; laevorotatory; mean free path; equivalent ionic conductance; liquid
L Avogadro constant; latent heat; luminance; angular momentum; geometric isomer of D form of a chemical compound; laevorotatory; leucine; left; Lambert
La lanthanum
LA left atrium; local anaesthetic
LAD language acquisition device, left axis deviation, left anterior descending (artery)
LAI leaf area index
LAN local area network
lat latitude
lb pound (weight)
lbf pound-force
lbf ft pound-force foot
lbf in^{-2} pound force per square inch
lbf ft^{-2} pound force per square foot
lb.s.t. static thrust in pounds

LBP low back pain, low blood pressure
lc lower case; liquid crystal
LC liquid chromatography
LCAO linear combination of atomic orbitals
LCD liquid crystal display
LCN load classification number
LD lethal dose
LD50 lethal dose 50%
LDH lactate dehydrogenase
LDL low density lipoprotein
Le Lewis function
LED light emitting diode
LEED low energy electron diffraction
Leu leucine
LF low frequency
LFER linear free energy relationship
LFT liver function test
LH luteinizing hormone
l.h. left hand
l.h.s. left hand side
LIF left iliac fossa
LIFO last in first out
LIH left inguinal hernia
lim limit
LIM-EMS Lotus-Intel-Microsoft Expanded Memory Specification
LIMS lazer ionization mass spectrometry
Li Lithium
liq liquid
LISP list processing
LKKS liver, kidney, kidney, spleen
LKS liver, spleen, kidney
LLL left lower lobe
LLQ left lower quadrant
LLTV low light television
lm lumen
LMFBR liquid metal cooled fast breeder reactor
LMP last menstrual period; left mento-posterior position of foetus
LMR liquid metal reactor
ln natural logarithm
LN lymph node
LNDO local neglect of differential overlap
LOA left occipito-anterior position of foetus
LOC loss of consciousness
LOCA loss of coolant accident
log logarithm to the base 10
long. longitude
LOP left occipito-posterior position of foetus
LP low pressure; lumbar puncture
LPG liquefied petroleum gas
LPH lipotropin
LPS lipopolysaccharide
LRH, LHRH, LRF luteinizing hormone-releasing hormone
LSCS lower segment caesarean section
LSD lysergic acid diethylamide
LSE London School of Economics
LSI large-scale integration
LST local sidereal time
LT local time
LTE local thermodynamic equilibrium
LTH luteotropic hormone (prolactin)
Lu lutetium
LUMO lowest unoccupied molecular orbital.
LUQ left upper quadrant
lut lutidine
LV left ventricle
LVET left ventricular ejection time
LVF left ventricular failure
LVH Left ventricular hypertrophy
Lw lawrencium
LWR light water reactor
lx lux
ly Langley
l.y. light year
Lys lysine

M

m mile; metre; million; minute; milli-; electromagnetic moment; molal; metastable;
magnification linear; mass; Earth's equivalent dipole; magnetic dipole moment; angular momentum component; meta (chemical descriptor); metal (in chemical formulas)
m- meta-
m_B molality

m_e mass of an electron
m_E mass of the earth
m_n mass of neutron
m_p mass of proton
M mega-; molar; magnetization per unit volume; magnifying power; moment of force; mutual inductance; torque; angular momentum component
Ma Mach number
M_B molar mass of compound B
mac macintosh
M&C(S) Microscopy and culture sensitivity of bacteria
MAD mutual assured destruction; magnetic anomaly detection
MADGE microwave aircraft digital guidance equipment
Man Mannose
Mane in the morning
MAO monoamine oxidase
MAR microanalytical reagent; memory address register
MASNMR magic angle spinning nuclear magnetic resonance
MATV master antenna television
max maximum
MB megabyte
MBC maximal breathing capacity
MBE molecular beam epitaxy
MBR memory buffer register
MBPT many body perturbation theory
MC/mc metric carat; Monte Carlo (method)
MCD magnetic circular dichroism
MCH mean corpuscular haemoglobin
MCHC mean corpuscular haemoglobin concentration
mCi millicurie
MCL monoclinic, mid-clavicular line
MCPA/MCP 4-chloro-2-methyl-phenoxyacetic acid
MCPB 4-(4-chloro-2-methyl-phenoxy)butanoic acid
MCPF modified couple pair functional
MCSCF multi-configurational self-consistent field
M_{crit} critical Mach number
Mc/s megacycles per second
MCV mean corpuscular volume
Md mendelevium
MD molecular dynamics
MDF main distribution frame
MDM mid-diastolic murmur
MDMA methylenedioxymeth-amphetamine (Informally "ecstasy")
MDA monochrome display adapter
M&E music and effects
Me methyl
ME myalgic encephalomyelitis
MEK methyl ethyl ketone (butan-2-one)
MEP molecular electrostatic potential
mEq milliequivalents
MERP minimum energy reaction path
Mes mesityl
MESFET metal-semiconductor field-effect transistor
meso- middle/intermediate
Met methionine; metastases
MeV mega-electron volts
M/F male/female
MF medium frequency; machine finished; molecular formula
Mg magnesium
mg milligram
MHC major histocompatibility complex
MHD magnetohydrodynamics
MHz megahertz
mho reciprocal ohm/unit of conductance
mi mile
MI mitral incompetence; myocardial infarction
MIBK isobutyl methyl ketone
MICR magnetic ink character recognition
μF microfarad
MIF migration inhibition factor
MIG metal inert gas welding
Mil-Spec military specification
MIM molecules-in-molecules
min minute; minimum; minim
MINDO modified intermediate neglect of differential overlap
mini- miniature/very small
MIPS millions of instructions per second
MIR mid-infrared
MIS Mullerian inhibiting substance

misc miscible
MISD multiple instruction stream single data stream
MIT Massachusetts Institute of Technology; monoiodotyrosine
mite/mitte dispense this number of tablets
mks metre-kilogram-second
MKSA metre-kilogram-second-ampere
ml millilitre
MLD mean lethal dose
Mn manganese
mm millimeter
MM molecular mechanics
MMA monomethyl aniline, N-methyl-aminobenzene
mmf micromicrofarad; magnet-omotive force
MMR mass miniature radiography; measles mumps rubella (vaccine)
Mo molybdenum
MOD Ministry of Defence
mol mole
mol. wt. molecular weight
mon monomeric form
MOP medical out-patient
MOPEG 3-methoxy-4-hydroxyphenylglycol
MOS metal-oxide semiconductor
MOSFET Metal-Oxide Semiconductor Field-Effect Transistor
MOU memorandum of understanding
m.p./mp melting point
MPa megapascal
Mpc megaparsec
MPD magneto-plasmadynamic; maximum permissible
MPEG/mpeg Moving Picture Experts Group
m.p.g./mpg miles per gallon
m.p.h./mph miles per hour
M.Phil. Master of Philosophy
MPR mannose-6-phosphate receptor
MPS Member of the Pharmaceutical Society
MQ metol-quinol
$\mathbf{M_r}$ Relative Molecular Mass
m/r modified release
MRC Medical Research Council
MRI magnetic resonance imaging
mRNA messenger ribonucleic acid
ms millisecond
MS multiple sclerosis; mass spectrometer; Master of Science; military specification, mitral stenosis, musculoskeletal
MS-DOS microsoft disk operating system
MSE mental state exam
MSH melanocyte stimulating hormone
MSL mean sea level
MSI medium scale integration
MSSU mid-stream specimen of urine
MSU mid-stream urine
MSW magnetic surface mail
MT 3-methoxytyramine
MTBF mean time between failures
M.Tech. Master of Technology
MTF modulation transfer function
MTI moving-target indicator
MTOC microtubule-organising centre
MUF maximum usable frequency
multi- many
Mur muramic acid
mV millivolts
MV megavolts; muzzle velocity
MVA motor vehicle accident
MVV maximal voluntary ventilation
MW megawatts; medium wave
MWE/MW megawatt electric
Mx maxwell
myo- muscle

N

n rotational frequency; number of moles; number of turns per unit length of coil; order of reflection /interference; principal quantum number; nano-; neutron; number density; refractive index, normal
N newton; nuclear; nitrogen; normality; neutron; north; number of mlecules; number of turns on a coil; modulus of rigidity; Avogadro number
$\mathbf{N_A}$ Avogadro number
Na sodium
$\mathbf{Na_E}$ exchangeable body sodium

N_A Avogadro constant
NA not applicable
NAA neutron activation analysis
NAD Nicotinamide adenine dinucleotide, No abnormality detected
NADH reduced nicotinamide adenine dinucleotide
$NADP^+$ Nicotinamide adenine dinucleotide phosphate
NADPH reduced nicotinamide adenine dinucleotide phosphate
nano- 10^{-9}
NAO natural atomic orbital
NART Nelson Adult Reading Test
NASA National Aeronautics and Space Administration
NASDA National Space Development Agency
Nb niobium
NBI no bone injury
NBO natural bond order
nbp normal boiling point
NBS N-bromosuccinimide
NCR no carbon required
Nd neodymium
ND normal delivery
NDB non-directional beacon
NDT non-destructive testing
N_E density of states
Ne neon
NE not engaged
NEFA nonesterified free fatty acid
neg. negative
NERC Natural Environment Research Council
Neu neuraminic acid
NEXAFS near-edge x-ray absorption fine structure
NFR not for resuscitation
ng nanogram
NGC New General Catalogue
NGF nerve growth factor
NGTE National Gas Turbine Establishment
Ni nickel
NIH National Institute of Health
NII Nuclear Installations Inspectorate
NIR near infrared
NIREX Nuclear Industry Radioactive Waste Executive
NK cell natural killer cell
nm nanometre
n.m./n mile nautical mile
NMR Nuclear Magnetic Resonance
NMRI Nuclear Magnetic Resonance Imaging
NND neo-natal death
No nobelium
No/Nos number/numbers
NOAA National Oceanic and Atmospheric Administration
nocte at night
Nonidet nonionic detergent
no-ops non-operable instruction
NOR nucleolar-organising region
Np neptunium
NP not palpable; nasal passage
NPH insulin neutral protamine Hagedorn insulin
NPK nitrogen, phosphorus and potassium fertiliser
NPL National Physics Laboratory
NPN nonprotein nitrogen
NPO nothing by mouth
NPU not passed urine
NPY Neuropeptide Y
NREM sleep nonrapid eye movement sleep
NRME notched, returned and mitred ends
NS nervous system
NSA no significant abnormality
NSAID non-steroidal anti-inflammatory drugs
NSF National Science Foundation
NSILA nonsuppressible insulin-like activity
NSILP nonsuppressible insulin-like protein
nt nit
NTG nitroglycerin
NTI noise transmission impairment
NTP normal temperature and pressure
NTS nucleus of the tractus solarius
NTSB National Transportation Safety Board
NTSC US National Television System Committee
NW narrow widths
NYD not yet diagnosed

O

O oxygen
o- ortho-
OA on admission; osteoarthritis
OAP old age pensioner
OB outside broadcast
OBM ordnance benchmark
Obs obstetrics
OC oral contraceptive
OCR optical character recognition
oct-/octa-/octo- eight
OCT orthoptic cardiac transplant
o.d. outer diameter, daily
OD right eye
ODA office document architecture
Oe oersted
O/E on examination
oed oedema
OEM original equipment manufacturer
OFHC oxygen-free high conductivity copper
OGD oesophago-gastro-duodenoscopy
OGF ovarian growth factor
om in the morning
OM otitis media
ONERA Office National d'Etudes et de Recherches Aerospatiales
OOH out of hours
OPD outpatient department
OPF orbiter processing facility
OPP film oriented polypropene film
OR operational requirement; operational research
ORB omnidirectional radio beacon
ORH orthorhombic
ORIF open reduction and internal fixation of fracture
Os osmium
OS Ordnance survey; one side; left eye
OS&W oak, sunk and weathered
osm osmole
OT occupational therapist
otc over the counter
OTU operational taxonomic units
OU both eyes
oz./oz ounce

P

p pico- ; pressure; penny/pence; impulse; momentum; atomic orbital "principal"; proton; electric dipole moment
p- para-
P phosphorus; physics; phase; electric polarisation; power; pressure; proline; partial pressure; pulse; protein
P_r / P_{fr} phytochrome
Pa protactinium; pascal
PA pernicious anaemia
PABX private automatic branch exchange
PAF platelet activating factor
PAH para-aminohippuric acid
PAL phenylalanine ammonia-lyase; phase alternation line
PAM pulmonary alveolar macrophage
PAR photosynthetically active radiation; precision-approach radar
para- beside
Par C parian cement
PAS para-aminosalicylic acid
PAT paroxysmal atrial tachycardia
PAX private automatic exchange
Pb lead
PBI protein bound iodine
PBX private branch exchange
pc after meals
PC personal computer, presenting complaint
PCB printed circuit board; post coital bleeding
PCC postcoital contraception
PCE pyrometric cone equivalent
PCM pulse coded modulation; pulse code modulation
PCNB quintozene
PCR polymerase chain reaction
Pd palladium
PDA patent ductus arteriosus
PDF portable document format
PDGF platelet derived growth factor
PDL page description language
pd/p.d. potential difference
PE pulmonary embolism
p.e.c. photoelectric cell
PEEP positive end expiratory pressure breathing

pel pixel
PEP pre-ejection period; phosphoenolpyruvate
PERLA pupils equal and reactive to light and accommodation
PERT programme evaluation and review technique
PET positron emission tomography, pre-eclamptic toxaemia
PETN pentaerythritol tetra-nitrate
PFK phosphofructokinase
PFR peak flow rate
PG prostoglandin
pH = $-\lg[H^+]$
PH past history
PhD/Ph.D. doctorate (Doctor of Philosophy)
Phe phenylalanine
pheno- derived from benzene
PHx past history
pi π
P_i inorganic phosphate
pica a unit of type size (1/6 inch)
PID personal identification device, prolapsed intervertebral disc, pelvic inflammatory disease
PIH, PIF prolactin-inhibiting hormone
PIN personal identification number
pk peck
pK = $-\lg[K_{eqm}]$
Pl plasma
PL/1 programming language 1
PLA programmable logic array
PLP pyridoxal phosphate
Pm promethium
PM purpose-made; powder metallurgy; postmortem
PMB postmenopausal bleeding
PMBX private manual branch exchange
PMC plaster-moulded cornice
PMH past medical history
PMN polymorphonuclear leukocyte
PMS pantone matching system
PMSF phenylmethylsulphonyl fluoride
PMX private manual exchange
PND paroxysmal nocturnal dyspnoea
PNdB perceived noise decibel
po by mouth
Po polonium
PO per orim
POCC payload operations control centre
POMC pro-opiomelanocortin
POP printing out paper; progesterone only pill; plaster of paris
POS terminal point of sale terminal
post-op postoperative
POY partially oriented yarn
PPC plain plaster cornice
PPH postpartum haemorrhage
ppi picks per inch
PP_i pyrophosphate
PPI plan position indicator
p.p.m./ppm parts per million
PPQ pterygopalatoquadrate
PPs peripheral pulses
ppt/ppte precipitate
pr by rectum
Pr praseodymium
PR Prothrombin ratio; per rectum
PRA plasma renin activity
PRBS pseudo random binary sequence
PRC plasma renin concentration
pre-op preoperative
PRF pulse repetition frequency
PRH/PRF prolactin releasing hormone
prn as needed, as required
Pro proline
PROLOG programming in logic
PROM programmable memory, premature rupture of membranes
$Prot^-$ protein
5-PRPP 5-phosphoribosylpyrophosphate
p.s.i./psi pounds per square inch
PSK phase shift keying
pt pint, patient
Pt platinum
PT Prothrombin time
PTA plasma thromboplastin antecedent
PTC plasma thromboplastin component
PTCA Percutaneous transluminal coronary angioplasty
Ptd.A. pointed arch

PTFCE polytrifluorochloroethene
PTFE polytetrafluoroethylene
PTH parathyroid hormone
PTS permanent threshold shift
PTSD post traumatic stress disorder
Pu plutonium
PU passed urine; peptic ulcer
PUO pyrexia of unknown or uncertain origin
pv by vagina/per vagina
PVA polyvinyl acetate
PVC polyvinyl chloride
PVD peripheral vascular disease
PVT paroxysmal ventricular tachycardia
PWR pressurised water reactor
PX private exchange
pyroGlu pyroglutamic acid
PZI protamine zinc insulin

Q

q banding techniques; heat
Q energy; symbol for charge; ubiquinone or plastoquinone; storage factor/quality factor/Q-factor/magnification factor
Q_{10} temperature coefficient
QA quality assurance
QAM quadrature amplitude modulation
qd every day
qds four times a day
qh every hour
QH_2 ubiquinol/plastoquinol
qid four times a day
qod every other day
QPP quiescent push-pull amplifier
QPSK quaternary phase-shift keying
qr quarter
QS quick sweep; quantity surveyor
QS_2 total electrochemical systole
QSO quasar
qt quart
quat quaternary ammonium salt

R

r specific refraction; population growth rate; racemic
r_E radius of the Earth
R/r radius
R resistance; Molar gas constant; rad/röntgen; arginine; prefix for structural configuration (right-handed); symbol for alkyl group; render; hydrocarbon radical; right; respiration
Rankine scale; Rydberg constant
R_H Hall coefficient
R_∞ Rydberg constant
Ra radium
RA rheumatoid arthritis
R.A. right ascension
rad radian
RAD right axis deviation
R&D Research and Development
RAE Royal Aircraft Establishment
RAeS Royal Aeronautical Society
RAIRS reflection-absorption infrared spectroscopy
RALS remote augmented lift system
Ram random access memory
RAM radar asorbing material, Relative Atomic Mass
RAS reticular activating system
RAST radioallergosorbent test
RAT ram-air turbine
RATOG Rocket Assisted Take Off Gear
Rb rubidium
RBC/rbc red blood cell (count)
RBL rhombohedral
RBS random blood sugar; Rutherford backscattering spectroscopy
RC rough cutting; reinforced concrete
RCA right coronary artery
RCS Royal College of Scientists; reaction control system
RDF radio direction finding
RDS respiratory distress syndrome
RDT&E Research, Development, Test and Evaluation
re rectus face
Re rhenium
RE resonance energy
RED radial electron distribution
ref refer
REF renal erythropoietic factor
reg regular
rem roentgen equivalent man

REM rapid eye movement; reflection electron microscopy
RER rough endoplasmic reticulum
rev revolution
r.f.\rf radiofrequency
Rf rutherfordium
RF radio frequency
RFI radiofrequency interference
RFLP restriction fragment length polymorphism
RFP reversed field pinch
RFS render, float and set
RFT renal function test
RGB red, green, blue
Rh rhesus factor; rhodium
r.h. right hand
Rha rhamnose
RHEED reflection high-energy elctron diffraction
RHF restricted Hartree-Fock (theory)
r.h.s. right hand side
RI respiratory infection
RIA radioimmunoassay
Rib ribose
Ribulo ribulose
RIF right iliac fossa
RIH right inguinal hernia
RISC reduced instruction set computer
RIST radioimmunosorbent test
RI strains recombinant imbred strains
RLL right lower lobe
RLQ right lower quadrant
R&M reliability and maintenance
rmm relative molecular mass
rms root mean square
RMV respiratory minute volume
Rn radon
RNA ribonucleic acid
RNAse ribonuclease
ROA right occipital anterior
ROHF restricted open shell Hartree-Fock
ROM\Rom read-only memory, range of motion
ROP right occipital posterior
RPA random phase approximation
RPF renal plasma flow
r.p.m./rpm revolutions per minute
RPN reverse polish notation
rps revolutions per second
RPV recoverable pilotless aircraft
RQ respiratory quotient
RR respiration rate
RRK Rice-Ramsperger-Kassel (theory)
RRKM Rice-Ramsperger-Kassel-Marcus (theory)
rRNA ribosomal ribonucleic acid
RRS resonance Raman spectroscopy
RS respiratory system
RSF rough sunk face
RSI repetitive strain injury
RSJ rolled steel joist
RTA road traffic accident
RTC real time clock, return to clinic
RTI respiratory tract infection
RTP/rtp room temperature and pressure
rT_3 3,3',5'-Triiodothyronine
Ru ruthenium
RUA right upper arm
Rubisco ribulose 1,5-bisphosphate carboxylase
RUBP ribulose 1,5-bisphosphate
RUDP ribulose 1,5-bisphosphate
RUQ right upper quadrant
RWP rainwater pipe
RV residual volume
RVE right ventricular enlargement
RVH right ventricular hypertrophy
Rx prescription, treatment
Ry rydberg

S

s distance along the path; second; svedberg; atomic orbital "sharp"; specific entropy; solid; conductivity; normal stress; Stefan-Boltzmann constant; surface charge density; surface tension; wave number; solubility; spin angular momentum; symmetry number; symmetrical
S sulphur; siemens; entropy; thermoelectric power; serine; svedberg; prefix for structural configuration (left handed); area; Poynting vector; probability current density; symmetry coordinate; spin angular momentum; supinator reflex

SA specific activity; surface area
SA node sinoatrial node
SAD Seasonal Affective Disorder
SAE Society of Automotive Engineers
SAH subarachnoid haemorrhage
SALC symmetry adapted linear combinations
SALI surface analysis by laser ionization
SALR saturated adiabatic lapse rate
SAM scanning Auger microscopy
SAN styrene and acrylonitrile
SANS small angle neutron scattering
Sar sarcosine
SAR synthetic aperture radar
SARAH search and rescue homing
sat, satd saturated
SAW surface acoustic wave
SAXS small angle x-ray scattering
Sb antimony
SB still-born
SBA standard beam approach
SBAC Society of British Aerospace Companies
SBC small bayonet cap
SBE sub-acute bacterial endocarditis
SBG serum blood glucose
sc subcutaneous
Sc scandium; Schmidt number
SCART Syndicat des Constructeurs d'Appareils Radiorécepteurs et Téléviseurs
Sc.D. Doctor of Science
SCE Scottish Certificate of Education; saturated calomel electrode
SCF self-consistent field
SCID severe combined immunodeficiency syndrome
SCPC single-channel per carrier
SCR silicon-controlled rectifier
scRNA small cytoplasmic ribonucleic acid
sd standard deviation
S.D./sd standard deviation
SDD subscriber direct dialling
SDI strategic defence initiative
SDS sodium dodecyl sulphate
Se selenium
S.E. standard error; stopped end
sec secant
sec- secondary
SECAM sequential couleur a memoire
SEELFS surface sensitive energy loss fine structure
SEM/S.E.M. standard error of the mean; scanning electron microscope
sepn separation
Ser serine
SER Smooth endoplasmic reticulum
SERC Science and Engineering Research Council
Serial input/output
SERS surface-enhanced Raman spectroscopy
SET single electron transfer
SEXAFS surface extended x-ray absorption fine structure
SG/sg specific gravity
SGHWR steam generating heavy water reactor
SGOT serum glutamicoxaloacetic transaminase
Sh Sherwood number
SH sulfhydryl, social history
Si silicon; sinister face
SI The International System of Units of Measurement.
SIMS secondary ion mass spectroscopy
sin sine
SINDO symmetrically orthogonalized INDO method
sinh hyperbolic sine
SIPN semi-interpenetrating polymer network
SISD single instruction stream, single data stream
SIT spontaneous ignition temperature
SkXR skull X-ray
sl sublingual
SL short lengths
SLAC Stanford Linear Accelerator Centre
SLAET Society of licensed Aircraft Engineers and Technologists
SLAM scanning laser acoustic microscopy
sln solution

SLR single-lens reflex; straight leg raising
Sm samarium
SM systolic murmur
SMC standard mean chord
SMO semiempirical molecular orbital
SMOW Standard Mean Ocean Water
SMR submucosal resection; standardized mortality ratio
Sn tin
S_N1 substitution nucleophilic unimolecular
S/N ratio signal/noise ratio
S_N2 substitution nucleophilic bimolecular
SNMS sputtered neutral mass spectroscopy
snRNA small nuclear ribonucleic acid
SNU solar neutrino unit
SO spin orbital
SOB short(ness) of breath
SOBOE shortness of breath on exertion
sol soluble; solution
soln solution
SOMO singly occupied molecular orbital
SOP surgical outpatients
Sor sorbose
SOS In an emergency
sp hybridisation of s and p atomic orbitals; species
sp^2 hybridisation of one s and two p atomic orbitals
sp^3 hybridisation of one s and three p atomic orbitals
SP soil pipe
SPCA proconvertin
SPECT single proton emission computed tomography
SPF/DB super plastic forming/diffusion bonding
sp. gr. specific gravity
SPM scanned probe microscopy
spp species
SPS super proton synchroton
sq square
SQ squint quoin
SQUID superconducting quantum interference device
sr steradian
Sr strontium
SR sinus rhythm
SRC Science Research Council
SRIF somatotropin release inhibiting factor
sRNA soluble ribonucleic acid
SROM spontaneous rupture of membranes
S-R theory stimulus and response theory
s/s signs and symptoms
SSI small scale integration
SS14 somatostatin 14
SSMS spark source mass spectroscopy
SST supersonic transport aircraft
SS28 somatostatin 28
St./st. stone (weight)
St stokes; Stanton number
ST surface trench
stat immediately
std standard
STD subscriber trunk dialing; sexually transmitted disease
STEM scanning transmission electron microscope
STH somatotropin/growth hormone
STI sexually transmitted infection
STM scanning tunneling microscopy
STO Slater type orbital
STOL short take-off and landing
STOVL short take off and vertical landing
STP/stp Standard Temperature and Pressure
STS space transportation system
sub sublimation, sublimes
SWG standard wire gauge
Sv sievert
SVC superior vena cava
SVD spontaneous vertex delivery; spontaneous vaginal delivery
sw swab
SWD short wave diathermy
sx/Sx symptoms
syn- same face

T

t time; transport number; tonne/ton; triton; thickness
T tera- ; tesla; tritium; temperature; moment of couple; period; thymine; transport number; kinetic energy; period; term value; torque; transmittance; triceps reflex
$T_{1/2}$, $t_{1/2}$ halflife
T3, T_3 3,5,3'-triiodothyronine
T4, T_4 thyroxine
Ta tantalum
T&A tonsils and adenoids
tab tablet
Tag tagatose
Tal talose
tan tangent
tanh hyperbolic tangent
TAPPI Technical Association of the Pulp and Paper Industry (US)
TAS true air speed
TASI time-assignment speech interpolation
TAT thematic apperception test
TATT tired all the time
Tb terbium
TB tuberculosis
TBC time-base corrector
TBG thyroxine binding globulin
TBO/tbo time between overhauls
TBPA thyroxine binding pre-albumin
TBW total body water
Tc technetium
TCA trichloroacetic
TCE trichloroethylene
TCGF T-cell growth factor
TCI to come in date
TCL triclinic
TCNE tetracyanoethene
TCSCF two configuration self-consistent field
TDF testis determining factor
TDMA time-division multiple access
TDRSS tracking and data relay satellite system
tds three times daily
TdT terminal desoxynucleotidyl transferase
Te tellurium
TE transverse electric
TEA tetraethylammonium
TED transmission electron diffraction, transferred electron device
TEM transmission electron microscopy; transverse electromagnetic
temp temperature
ter- three
tera- 10^{12}
tert- tertiary
TET tetragonal
TETRAC tetraiodothyroacetic acid
TE-wave transverse electric wave
Tf trifluoromethanesulphonyl group
TFD Thomas-Fermi-Dirac (method)
TFTR tokamak fusion test reactor
TFTs thyroid function tests
TGA thermo-gravimetric analysis
Th thorium
theor theoretical
thf/THF tetrahydrofuran
THI temperature-humidity index
THORP thermal oxide reprocessing plant
THP tetrahydropyranyl group
Thr threonine
THR total hip replacement
Thy thymine
Ti titanium
TI tricuspid incompetence
TIA transient ischaemic attack
tid three times daily
TKR total knee replacement
Tl thallium
TL transmission loss; thermoluminescence
TLC thin layer chromatography; total lung capacity
Tm thulium; renal tubular maximum; melting temperature
TM transverse magnetic; tympanic membrane
TMJ temporo-mandibular joint
TMS tetramethylsilane
tn ton
TNS transcutaneous nerve stimulator
TNT trinitrotoluene
Tol tolyl
TOP termination of pregnancy
Torr torr
TP turning point

tPA tissue plasminogen activator
TPHA treponema pallidum haemagglutination
TPI threads per inch; tons per inch
TPN triphosphopyridine nucleotide
TPNH reduced triphosphopyridinenucleotide
TPP thiamine pyrophosphate
TPR total peripheral resistance, temperature pulse respiration
TR temporary resident
TRACE test-equipment for rapid automatic checkout and evaluation.
TRAPATT diode trapped plasma avalanche transit-time diode
TRG trigonal
TRH/TRF thyrotropin releasing hormone
tri- three
TRITC tetramethyl rhodamine isothiocyanate
tRNA transfer ribonucleic acid
Trp/Try/Tryp tryptophan
trs transition
TS tricuspid stenosis; transition state
Ts&As tonsillectomy and adenoidectomy
TSH thyroid stimulating hormone
TShR total shoulder replacement
TSI thyroid stimulating immunoglobulins
tsp teaspoon
TT tetanus toxoid; tuberculin tested
TTL transistor-transistor logic; through the lens
TTO to take home/away (medication)
TTP thymidine triphosphate
TTS temporary threshold shift
TTX tetrodotoxin
TU traffic unit; transmission unit
TURP transurethral prostate resection
TV trichomonas vaginalis; tidal volume
TVP textured vegetable protein
2,4,5,-T 2,4,5,-trichloro-phenoxyethanoic acid
Tyr tyrosine

U

u object distance; atomic mass unit (unified); velocity; specific internal energy; Bloch function; electric mobility
U uranium; internal energy; uracil; unit; tension; electric potential difference, urea
U and E urea and electrolytes
UART universal asynchronous receiver transmitter
UC ulcerative colitis
UCR unconditioned reflex
UCS unconditioned stimulus
UDP uridine diphosphate
UDPG uridine diphosphoglucose
U&Es urea and electrolytes
UF urea-formaldehyde
UFA unesterified free fatty acid
UFO unidentified flying object
UGS urogenital system
UHF ultra high frequency; unrestricted Hartree-Fock
UHT ultra heat treated
ULA uncommitted logic array
UMN upper motor neurone
UMP uridine monophosphate
uns/unsym unsymmetrical
UPES ultraviolet photoelectron spectroscopy
UPS ultraviolet photoelectron spectroscopy
ur urea
Ura uracil
URF uterine relaxing factor/relaxin
URTI upper respiratory tract infection
USP United States Pharmacopeia
USS ultrasound scan
UT universal time
UTI urinary tract infection
UTC universal time coordinates
UTP uridine triphosphate
UV/u.v./uv ultra-violet (spectroscopy)
UVL ultra-violet light

V

v image distance; velocity; specific volume of a gas; reaction rate; vibrational quantum number; vicinal;

V volt; volume; electric potential; vanadium; valine
V_m molar volume
VAB vehicle assembly building
Val Valine
VAM vesicular-arbuscular mycorrhiza
vap vaporization
VAr volt-amperes reactive
VASI visual approach slope indicator
VB valence band; valence bond
VC vital capacity
VCD vibrational circular dichroism
VCO voltage controlled oscillator
VCR video cassette recorder
VD venereal disease
VDISK virtual disk
VDRL venereal disease reference laboratories / venereal disease research laboratory
VDU visual display unit (monitor)
VE vaginal examination
VEGF vascular endothelial growth factor
VF ventricular fibrillation
VFR visual flight rules
VGA Video Graphics Array
VHF very high frequency
VI virgo intacta
vic vicinal
VIP vasoactive intestinal polypeptide
Vit/vit vitamin; vitreous
VLBI very long baseline inferometry
VLDL very low density lipoprotein
VLF very low frequency
VLS vapour-liquid-solid
VLSI very large scale integration
VMA vanillylmandelic acid
VMC visual meteorological conditions
VOR vestibulo-ocular reflex
VP vent pipe, venous pressure
VPI vapour phase inhibitor
vs versus, vital signs
VS volumetric solution
VSEPR valence shell electron pair repulsion
VSI vertical speed indicator
VSLI very large scale integrated circuit
VSTOP vaginal suction termination of pregnancy
VSWR voltage standing-wave ratio
VT ventricular tachycardia
VTR video tape recorder
VU volume unit
VUV vacuum ultraviolet
v/v volume per volume

W

w work; energy density; mass fraction; work
W energy; watt; tungsten; tryptophan; electrical energy; weight; radiant energy; statistical weight; work
WAIS Wechsler Adult Intelligence Scale
WAT word association test
WAXS wide angle x-ray scattering
Wb weber
WBC/wbc white blood cell
WCC white cell count
We Weber number
wf wrong font
win windows
WISC Wechsler Intelligence Scale for Children
WKB Wentzel-Kramers-Brillouin (method)
WNL within normal limits
wp word processor/word processing
w.p.m./wpm words per minute
WR Wassermann reaction
wrt with respect to
wt weight
w/v weight per volume
w/w weight per weight

X

xB (or **yB**) mole fraction of substance B
X reactance
XAFS x-ray absorption fine structure
XANES x-ray absorption near-edge structure
Xe xenon
XPES x-ray photoelectron spectroscopy
XPS x-ray photoelectron spectroscopy
XR x-ray
XRD x-ray diffraction

XRF x-ray fluorescence
XRS x-ray spectroscopy
Xu X unit
Xyl xylose
x,y,z Cartesian coordinates

Y

y./y/yr year
Y yttrium; tyrosine; admittance
Yb ytterbium
yd./yd yard
YIG yttrium iron garnet
YOB year of birth
yr year

Z

z charge number of an ion; collision frequency factor
Z atomic number; impedance; Earth's vertical magnetic field; on the same side; number of molecular collisions per second; section modulus; compression factor; collision number; partition function; zusammen
ZDO zero differential overlap
z.f. zero frequency
Zn zinc
ZPE, ZPVE zero point vibrational energy
Zr zirconium
ZULU Greenwich mean time

ANSWERS

Unit 1: Phrasal Verbs (Act - Break)

A)
1) He was awake all night preparing his presentation.
2) This is a reliable piece of equipment that has never failed.
3) The useful information in the lecture could be summarized in a few sentences.
4) This stopwatch is working incorrectly (*=is not working correctly*), could I borrow yours?
5) The results were unexpected and difficult to explain.
6) The engine must be tested until it fails.
7) Did you think that what he said was logical?
8) We were unable to explain the patient's temperature.

B)
1) This pH meter is acting up - again!
2) How do you account for the physical properties of the polymer?
3) I did not realize that he was in on the development of the IBM PC.
4) Is it possible to break through rapidly in this area of research?
5) He was so excited by the work that he was awake all night so that it would be finished on time.
6) The meeting was a waste of time; all the useful information could be boiled down to one page.
7) The patient's presenting complaint was boiled down in(to) three words; severe abdominal pain.
8) They were not involved in the randomized controlled trial, so they knew nothing about the preliminary results.

Unit 2: Phrasal Verbs (Break - Come)

A)
1) The carnivore carries on chasing the herbivore until the herbivore is captured.
2) He enjoys catching out bad speakers at conferences; he wants to appear clever.
3) The geological expedition will begin two weeks after the term/semester breaks up.
4) We expect that the cost of software will come down over the next five years.
5) I really should find time to catch up on the new genetic technology.
6) The operator can bring down the temperature of the nuclear reactor.
7) How many publications have they brought out this year?
8) While reading Nature I came across an interesting article.

B)
1) Mendax pharmaceuticals have *published* a new catalogue. *Or*
   Mendax pharmaceuticals have *produced* a new catalogue.
2) We must decrease the temperature of the furnace by 200°C. *Or*
   We must reduce the temperature of the furnace by 200°C
3) I can't continue this line of research without further funding.
4) Although the speaker was well prepared, he was outwitted by penetrating questions.
5) I need to review the latest developments in solid state physics.
6) During the survey we noticed three new species.
7) During the autumn the temperature and humidity *decrease* a little. *Or*
   During the autumn the temperature and humidity *fall* a little.
8) Continue giving the patient intravenous antibiotics for two more days.
9) We need to *reduce* your blood pressure to 130/90 mmHg. *Or*
   We need to *decrease* your blood pressure to 130/90 mmHg.

Unit 3: Phrasal Verbs (Come - Die)

A)
1) I described the problem and she instantly came out with five solutions. *(The implication is that she produced the solutions verbally.)*
2) After coming up with a brilliant idea he shouted "Eureka".
3) I think that we can count on him to successfully complete the project.
4) I was fixing the VDU (screen) when another problem with the computer cropped up.
5) In order for our cars to be ecologically acceptable we must cut down on (*or cut back*) toxic emissions.
6) The resistance of electric cables can be cut down (*or cut back*) by 90% by using superconductors.
7) The fluorescence will die down within nanoseconds.

8) Did anything crop up during your neurological examination of Mrs Smith?

B)
1) During the discussion he produced a new idea for a solar powered house.
2) That compound is likely to be under the heading of hydrocarbons. *Or*
That compound is likely to be classified under the heading of hydrocarbons. *Or*
That compound is likely to be categorized under the heading of hydrocarbons.
3) Kekulé produced the structure of benzene after dreaming about a snake.
4) Can we rely on the Romega II sales representative to tell us the truth? *Or*
Can we depend on the Romega II sales representative to tell us the truth?
5) If no complications occur unexpectedly then the rocket should arrive at the moon before 12 pm. *Or*
If no complications happen unexpectedly then the rocket should arrive at the moon before 12 pm.
6) If we want the research budget to last for the whole year, we must hold back spending now.
7) Decreasing fat consumption decreases your risk of cardiovascular disease.
8) The controversy about genetic technology is decreasing (*or diminishing*).
9) You cannot always rely on the pulse oximeter's measurement of oxygen saturation.
10) During the multidisciplinary team meeting he produced a brilliant diagnosis.

Unit 4: Phrasal Verbs (Do - Get)

A)
1) Is it wise for this country to do without nuclear power?
2) We cannot afford to drop out of the race to develop a popular electric car.
3) The farmers in this area should ease up on their use of nitrates.
4) The Europeans are falling behind the Americans in the race to explore outer space.
5) If you don't feed the paper into the photocopier correctly, it won't work.
6) These results don't fit in well with your hypothesis; you should repeat the experiment.
7) Can you figure out the solubility of this compound from its solubility product?
8) It is wise to follow up interesting observations because they may reveal something important.
9) The least interesting part of any clinical trial is feeding the data into the statistics package.
10) The hospital cannot do without an intensive care unit.

B)
1) If you want to finish the assignment before April, you should not slow down now.
2) If you make little progress on your assignments you will have to work very hard over the vacation.
3) Key in the data by hand; there is no automatic transfer between the two computers. *Or*
Type in the data by hand; there is no automatic transfer between the two computers.
4) A scientist with your training and experience would be compatible with this research group.
5) It shouldn't take long to deduce a way to make this organic compound. *Or*
It shouldn't take long to work out a way to make this organic compound.
6) Her seminars are well attended because she is a good speaker who communicates her message well.
7) He is often invited to attend conferences because he has good relationships with his peers.
8) Your morbidity and mortality statistics show that your performance is second to the other surgeons.
9) 20% of the doctors will abandon this very competitive postgraduate course. *Or*
20% of the doctors will give up this very competitive postgraduate course. *Or*
20% of the doctors will stop doing this very competitive postgraduate course.

Unit 5: Phrasal Verbs (Get)

A)
1) A university researcher cannot get by with only one publication per year!
2) The meteorologist hoped that the bad weather would not get us down.
3) I tried to get down the information from the slide before the speaker replaced it.
4) Stop talking and get on with the test.
5) He's the sort of mathematician who can get out an answer to any problem.
6) Quantum mechanics is not an easy subject to get over to the general public.
7) When will you get round to writing the review article?
8) I can't get away with showing results that are not statistically valid.
9) She has a post-natal depression that it makes it difficult for her to get on with her usual way of life.

B)
1) The Research Council has decreased our funding by 20%; we must manage with less money.
2) I'm sorry I haven't made time to calculate the mean and standard error of these data yet.
3) He didn't communicate that point very well in his seminar.
4) Let's continue designing the new car - after lunch!
5) He applied for many computing jobs unsuccessfully but did not let rejection depress him.

6) When I first looked at the problem I didn't think I'd be able to work out a solution.
7) Can you avoid the consequences of dissolving the protein in ethanol? Won't it be denatured (damaged)?
8) The Professor of Urology wants to discharge his clinical responsibilities as soon as possible and return to research.
9) A paediatrician cannot avoid the consequences of a poor bedside manner for long.

Unit 6: Phrasal Verbs (Get - Go)

A)
1) Will you be able to get through counting all the animal species in your quadrant before lunch?
2) Everbody must give in their practical books by 4pm.
3) They have given up (on) their attempt to find a polymer as strong as steel.
4) If you go by this thermometer then the temperature in this room is -20°C!
5) Go ahead with the experiments that were suggested by Masters and Johnson.
6) Could you go into the background of your research in more detail?
7) Please go into the daily responsibilities of my new position as the lead gastroenterologist.

B)
1) How long will it take to complete the isolation and purification stages of this protein?
2) I will have to finish and submit the report on drug toxicity before Thursday.
3) We should abandon this approach and try to find another way to synthesize this antibiotic.
4) Judging by the reading on this gauge the pressure is 60 pounds per square inch.
5) If you judge by the pole star, north is over there.
6) Now that we have the funding we can proceed with the trip to the Artic.
7) In the second half of the seminar he described the machine's components.
8) I must make an urgent call but the departmental telephone network has gone down..
9) He could tell by the silence that the lecture had not been well received.
10) Every doctor in the UK must complete their membership exams before they can become a consultant.

Unit 7: Phrasal Verbs (Go - Keep)

A)
1) If you store the reagent at room temperature it will go off in a week.
2) The company has decided not to go on with the development of laser weapons.
3) Please go over your strategy for developing a cure for AIDS.
4) There is a rumour going around that you may win this year's Nobel prize.
5) The number of species in this area has gone up by 15%; so increasing biodiversity.
6) By the end of the year I should have three more articles out in Nature.
7) Could you go back to the library and hunt down this reference?
8) We must keep to our plan to build a robotic surgeon within five years.

B)
1) Is there gold in this soil or should I ignore the rumour circulating?
2) The cost of the Scientific American has increased by 10%.
3) We are planning to publish the article by March.
4) I had no difficulty finding the telephone number of the Science Research Council.
5) The department will be permanently closed if we do not follow the safety regulations.
6) The Ferrari does not have enough power to stay with the Lotus; the Lotus will win.
7) A literature search should help you find the important article.
8) The number of people applying for fellowships in Medical Education has gone up by 35%.
9) Both the employer and employee must adhere to the terms of the contract.

Unit 8: Phrasal Verbs (Key - Look)

A)
1) To get money from the machine you must key in your personal identification number (PIN).
2) The car's exhaust lets toxic gases out into the atmosphere.
3) If you look at the cell under the microscope you can see several nuclei.
4) We are looking forward to the results of the drug tests on humans.
5) Maybe you should look into the possibility of cloning this gene.
6) I gave the article to the professor to look over before I submitted it to Nature.
7) I can never remember the exact speed of light, I always have to look up the precise value.
8) Dermatologists need to look at skin lesions carefully and describe them accurately.

B)
1) I hope the development of voice activated computers means that I will not have to type in information.
2) If you turn on the tap you will release the water at a pressure of 60 lbs per square inch.
3) If you examine the sun through a telescope, you may be blinded.
4) I (optimistically) anticipate the result of the organic synthesis; I am sure that the yield will be high.
5) I am investigating the possibility of buying a portable PC (personal computer).
6) After examining his mathematical proof for the third time, I found his mistake.
7) Search for the hyperkalaemia protocol in the acute medicine handbook.
8) Because he has won the Nobel prize for Chemistry, many people admire him.
9) I am (optimistically) anticipating a percentage yield of at least 80%.
10) Have you investigated the possibility that the disease is a vasculitis rather than a lymphoma?

Unit 9: Phrasal Verbs (Make - Put)

A)
1) Make up a 5% solution of $NaHCO_3$ in water.
2) If the telescope's resolution is not good you will not be able to make out adjacent stars.
3) I never pick up everything in a discussion about genetic technology.
4) The pharmacist will take 10 minutes to make up your medicine (=drug).
5) Take 2 mls of the supernatant and make it up to 5 mls with water.
6) Good scientific communication will prevent people in the audience from nodding off!
7) The government put the increase in cancer deaths down to poor nutrition.
8) The weighing scale must be playing up again, I don't believe its readings.
9) His success in academic medicine was put down to his high IQ.

B)
1) To make a threefold dilution, take 1 ml of solution and increase it to 3 mls with the solvent.
2) If you look into the distance you can see one of our planes.
3) Because I was carrying out an experiment late last night, I went to sleep during today's physics lecture.
4) Could you repeat the explanation, I did not understand everything you said.
5) If the centrifuge starts to work incorrectly call the repairman.
6) This satellite project will fail; it's time to withdraw and leave it to the Americans.
7) The investigators attributed the plane crash to a mistake by the pilot.
8) This patient has three first degree relatives who have had myocardial infarctions under the age of 60 – we attributed this to a family history of hypercholesterolaemia.
9) The spectrometer is malfunctioning again. I think it is time to buy a new one.

Unit 10: Phrasal Verbs (Put - Reel)

A)
1) I have put in for a grant from the National Institute of Health.
2) The government put off making a decision about the drug until new safety tests were done.
3) Put on your lab coat!
4) Put out the ultraviolet light before you enter the room.
5) Operator, could you put me through to the head office of BRITCO Pharmaceuticals?
6) Many people believe that we are reading too much into the loss of the ozone layer.
7) Because she had read up on protein purification techniques she prepared an excellent protocol.
8) Her complete understanding of her field allowed her to reel off facts confidently.
9) The statistician does not believe that our data is significant at the $P<0.05$ level; it looks as though we were reading too much into the data.

B)
1) I have applied for the job of Director of Research and Development.
2) I shall delay applying for a job as a geologist until I have got my doctorate.
3) The streetlights are turned off automatically at dawn each day.
4) If Professor Bell's secretary does not put me through to him, I can contact him by e-mail.
5) He told me that I was overinterpreting my data - I needed to do more experiments.
6) The salesman confidently repeated the specifications of the music system.
7) Turn the colorimeter on 30 minutes before you use it.
8) Having reviewed the physics of subatomic particles, I was able to make far more intelligent suggestions during my conversation with him.
9) After reviewed the pathology of liver diseases, I was able to learn far more during my attachment in medical gastroenterology.

Unit 11: Phrasal Verbs (Round - Run)

A)
1) I ran across the Safety Officer in the library; he reminded me of the new regulations.
2) The DNA fingerprint evidence was enough to rule him out as a suspect.
3) He rounded off the article with a section on possible future technological developments.
4) Run the chart recorder at 12 cm/min to get a satisfactory trace (graphical record).
5) Could you run off 7 copies of this diagram for me?
6) We've run out of butane, so the gas cylinder is empty.
7) There are one or two things I am not clear about, can we run through them again?
8) Can we run over the risk factors for atherosclerosis?
9) The most recent experimental data rule out your favoured hypothesis.
10) It would be helpful if you could run through your proposed theory again. I didn't understand it the first time.

B)
1) She finished the microbiological conference with a summary of recent discoveries.
2) We cannot exclude the possibility of electric polymers being used to make new TV (television) screens.
3) It was during the Marine Biology Conference that she met her previous supervisor.
4) If you operate the centrifuge at 2000 revolutions per minute you will separate the components.
5) The professor's secretary made 5 copies of the grant application.
6) I have used up the space on the hard disk.
7) Let's review this report before the meeting starts.

Unit 12: Phrasal Verbs (Run - Slip)

A)
1) He screwed up the computer after accidentally pouring coffee on the keyboard.
2) He was sent down from university for stealing drugs from the pharmacology department.
3) I can't buy a Geiger counter here, I'll have to send off for one.
4) That surprising result set back our work for 8 months.
5) Five years ago the physics department set up a project to develop an electronic nose.
6) The animal can change colour so that it does not show up against the background.
7) I am less likely to slip up using a Windows based program than an MSDOS based program.
8) All flights from the airport are cancelled until the weather settles down.
9) The junior doctor discovered that his job applications were being set back by a poor reference.
10) I have just sent off for the latest edition of Gray's Anatomy.
11) After taking the antacid tablets, her abdominal pain settled down.

B)
1) It will take 2 weeks to make a miniature model of the plane and then test it.
2) Cigarettes wreck your lungs and circulatory system.
3) Is it better to get a bad degree classification than to be expelled?
4) After seeing the advertisement for a toxicologist I wrote for an application form.
5) Losing his data would have held him back for months but fortunately he had backed up a copy of the data.
6) The computer network is organized with a password system to protect the user's data.
7) It will take 4 hours to assemble the experiment but only 30 minutes to perform it.
8) The cardiology consultant made a mistake and missed a cardiac murmur.

Unit 13: Phrasal Verbs (Slow - Take)

A)
1) You must slow down your speech for the voice recognition system to work.
2) Could you spell out your plans for your research division over the next 5 years?
3) The water peak stands out in this spectrum.
4) He summed up his model of how the brain works.
5) As he described the Schrödinger equation, most of the students switched off!
6) If the speaker talks too quickly you will not have time to take down everything.
7) The latest modem took off last summer when sales increased dramatically.
8) The aspect of the presentation that stands out most, is the discussion of cancer prevention by dietary alterations.

B)
1) He speaks so clearly that it is very easy to absorb what he has said.
2) The electric field accelerates the electrons by 500 m/s.
3) The programming manual would be better if it clarified the instructions.

4) The text is more prominent (or *obvious*) on a light emitting diode display than on a liquid crystal display.
5) The science reporter wrote down every word that the Professor of Genetics said.
6) You must exit the program before turning off the computer.
7) The rocket decelerated to 5 m/s before it landed on Pluto.
8) He summarized the multimedia potential of the latest personal computer.
9) You have to switch off all mobile phones before entering the hospital; this is particularly important for the Intensive Care Unit.

Unit 14: Phrasal Verbs (Take - Wade)

A)
1) Put the glass equipment on the shelf and turn on the autoclave.
2) I turned down the velocity of the centrifuge to 3000 rpm (revolutions per minute).
3) The thermostat will turn off (or *turn down*) the heating if the house becomes too warm.
4) Virtual reality turned out to have more applications than first anticipated.
5) We will take on the task of designing and building the next Boeing plane.
6) I turned to my supervisor for technical advice.
7) Turn on the High Performance Liquid Chromatography machine and load the sample.
8) This hospital is taking on more patients than it can manage; we need to start redirecting patients to nearby hospitals.

B)
1) We calculate that the petroleum in this oil field will run out in 2 years.
2) As the reagent (reactant) is used up the chemical reaction decreases in speed.
3) It took me 6 hours to get through the review article on the management of asthma.
4) I arrived at the conference just in time to give my presentation.
5) Turn on the Blu-ray player and play the Region 1 disc.
6) I referred to the Annual Review of Physics for more information.
7) The printing press runs continuously; it is not switched off.
8) The lights were decreased in intensity (or *dimmed*) the projector was switched on and the film was shown.

Unit 15: Phrasal Verbs (Warm – Write)

A)
1) The tyres on the car warm up as the car speeds up.
2) A Japanese company has worked out a way to make a computer the size of a pea.
3) Organic chemists have worked out a way to synthesize the sugar.
4) I weighed up the alternatives before making a decision on where to drill for petroleum.
5) It does not take long to write up an abstract
6) She weighed up the advantages and disadvantages of each antipsychotic medication.
7) As the metal warms up, its electrical properties change.
8) I worked out the probability of rain as 70%.
9) I worked out the required dosage by allowing 1mg for each10kg of bodyweight.

B)
1) After you switch the machine on it must warm up for 2 minutes before you can use it.
2) Deduce a solution to the mathematical equation.
3) I assessed the consequences of staying there against starting my own software business.
4) Someone has deduced that the mercury in tooth fillings causes kidney damage.
5) As part of the grant application he had to write a report of the group's work.
6) The oven must heat up before it can be used.
7) The journalist wrote an account of his interview with Professor Crick.
8) He has deduced how to synthesize a compound more versatile than the Grignard Reagent.

Unit 16: Connectives - Addition (Answers)

A)
1) And
2) Also
3) In addition
4) And
5) Also, in addition, moreover *or* furthermore
6) And
7) In addition *or* also
10) Also
11) And

8) Furthermore *or* moreover
9) In addition

B)
1) Furthermore
2) Also
3) In addition
4) Moreover
5) Furthermore

Unit 17: Connectives - Deduction (Answers)

Scenario 1) A teacher talking to a student alone after class:
Therefore the compound must be ethene. *Or*
So the compound must be ethene.

Scenario 2) Extract of a discussion from a published article.
Thus they are more susceptible to heat loss than large animals. *Or*
Hence they are more susceptible to heat loss than large animals.

Scenario 3) A lecturer in nutrition talking to a general practitioner.
Either they are obese **or** they are very muscular athletes.

Scenario 4) Two students trying to understand some physics during their lunch break.
So if the mass of the object is constant and gravity is constant, then the object's weight must be constant.

Scenario 5) Two inexperienced software developers.
Is there a logic function that requires **neither** A nor B to be true to trigger an output?

Scenario 6) A city planning office.
At that junction the drivers have a choice between taking a short distance route **or** taking a longer distance route that is actually faster. **So** most take the second route – leading to predictable traffic congestion.

Scenario 7) Sports commentator currently describing a race.
Neither the second nor the third place runner are even close to him **so** they have no chance of breaking **either** World **or** Olympic records.

Unit 18: Connectives – Cause (Answers)

A)
1) The rise in body temperature was **caused** by decreased conduction, convection and radiation.
2) The lifetime of this solar system is limited **because** of the nuclear reactions occurring in the sun.
3) On **account** of his acute and severe blood loss, the patient suffered multi-organ failure and died.
4) The **reason** for the decrease in the number of this species, is the increase in the number of its predators.
5) The **cause** of the patient's encephalitis was a herpes virus infection.
6) An object accelerates **because** of the resulting force acting on it.
7) Her cancer was **caused** by an inherited ras oncogene.

B)
Scenario 1) Furthermore I think that these headaches are, in fact, migraines because they seem to be **caused** by chocolate.
Scenario 2) I imagine it was **because** of their more efficient engine and a better drag coefficient.
Scenario 3) On **account** of the stagnant economy and poor growth figures, a change in the interest rates is likely. However some commentators believe that the **reason** for the change in interest rates is actually the government's concern about the approaching elections.
Scenario 4) **Because** of its structure, fewer steps are required to complete the synthesis. So it is a cheaper synthetic route.
Scenario 5) The **reason** for my purchase becomes clear if you consider the following.

Unit 19: Connectives – Contrast (Answers)

A)
1) Using V=IR we would expect the voltage drop across this component to be 6 mV; **however**, it is 12 mV.
2) "The operation was a success **but** the patient died." Maybe you need to reconsider what you call success! (Best answer)
"The operation was a success **yet** the patient died." Maybe you need to reconsider what you call success! (Acceptable answer)

3) Until recently tetrahydrides were being synthesized for use as superconductors. **However** (or **yet**), the latest research indicates that trihydrides can act as superconductors at lower more attainable pressures. This is directly in **contrast to** work done in our laboratory, which indicates that trihydrides are not good superconductors.
4) **Contrary to** expectations the temperature of the city fell more than that of the local countryside last Monday. Why didn't the urban heat island effect apply then?
5) On the one hand, we are under pressure to deliver good and economical patient services by using computers. **On the other hand** weare not being given the money, equipment or legislative support to ensure widespread use of computers in medicine.
6) This species was declared extinct ten years ago. **Yet** (or **But**) during our expedition in Chile we came across a freshly decomposing male of the species.
7) Currently, rapid wireless communications make use of radiowaves as the transmission medium. **However,** because radiowaves can pass through buildings, the information can be intercepted and so is not secure. In **contrast to** this, the latest optical broadband would be more secure because it is restricted to inside the building. Additionally, the optical technology has a greater bandwidth.
8) **Contrary to** expectations, long term oral ingestion of antibiotics actually predisposes an individual to secondary bacterial infections.
9) On the one hand patients with depression have an increased risk of mortality **on the other hand** patients with depression and anxiety have a lower mortality compared with depression alone.
10) **Contrary to** current standards, in the future all nuclear reactors will be based on fusion rather than on fission.

Unit 20: Connectives – Comment (Answers)

Scenario 1) A surgeon describing the results of his first attempt at a cardiac transplant to his boss.
The heart we used was harvested from a young donor in a persistent vegetative state. It was brought to our hospital by helicopter. I myself examined the heart to confirm that it was in good condition. The operation itself went very smoothly and only took three hours. The recipient was cardiovascularly stable through the procedure. There were no complications. ***Unfortunately,*** although the operation was a success, the patient died. We have requested an autopsy to find out why the patient died.

Scenario 2) A postdoctoral scientist talking to her husband at the end of a busy day.
It has been a tough day. I repeated the experiment 5 times. I tried changing all the reagents and most of the equipment and I still didn't get the result I was expecting. ***Obviously,*** I will try again tomorrow but I am not hopeful.

Scenario 3) Architect talking to a project manager.
The possible renewable energy sources we considered were wind energy, solar energy, geothermal energy and wave energy. Each has its strengths and weaknesses. ***Naturally,*** in a region with abundant sunlight throughout the year, solar energy is preferred. ***Obviously,*** in an area with less direct sunlight and more wind, wind energy would be a better choice.

Scenario 4) Botanist lecturing a student.
Photosynthesis requires light to convert carbon dioxide and water into carbohydrate and oxygen. Although this is the accepted dogma, ***actually*** not all the frequencies of visible light are necessary for this process. For example, green chlorophyll has a different light absorption spectrum from brown chlorophyll but they can both support photosynthesis.

Scenario 5) Human Resources Manager discussing job applications with a colleague over lunch.
We use multiple choice tests to assess the candidates and these results play a major role in the selection process. Is it appropriate for a written test to be so important in the selection process? Has this particular written test even been properly validated? ***Frankly,*** I believe strongly that the candidate's past performance in real work situations should be given greater consideration than a test.

Scenario 6) School chemistry teacher talking to his/her class.
Most people believe correctly that sulphuric acid is a strong acid. However non-scientists frequently confuse corrosiveness with strength. ***Actually,*** sulphuric acid is a strong acid because it dissociates freely in water and not because it is corrosive. ***Of course,*** acid strength correlates positively with corrosiveness.

Scenario 7) Extract from a journal article.
Pure water solidifies at about 0°C under normal atmospheric pressure. Similarly, ice melts at 0°C. As the temperature is raised, the energy of the water molecules increases until they can go from the liquid state to the gaseous state. ***Obviously*** then, the boiling point of any liquid will higher than the melting point of the same material.

Scenario 8) A forensic pathologist giving evidence at a trial.
Fingerprint comparison requires a twelve point match for identification. We recovered four fingerprints from the scene of the crime. ***Unfortunately,*** in this case we only have six point matches, so the identification is unreliable.

Unit 21: Connectives – Speculation (Answers)

A)
1) If you drive to Paris will your carbon footprint be less than taking a plane?
2) If your iodine intake is too high will you be prone to hyperthyroidism?
3) If you look directly at the sun through a telescope, will you be blinded?
4) If everybody on the planet were a vegetarian would there be enough food for everyone?
5) If sales of the New Scientist increased at that rate would they overtake the Scientific American?
6) If we lost the ozone layer, would life on the planet cease?

B)
1) If we **dig** beneath the layer of sedimentary rock, what do you think we will find?
2) If she were exposed to Lancefield Group A beta haemolytic streptococci in her current state, what **would happen**?
3) If we alter this part of the molecule, **will** it **be** effective at the receptor?
4) Would the pipes burst if we **increased** the gas pressure?
6) What **would happen** if the government were to cut public spending by 20%?
7) If you **increased** the carbon content of this steel would it be stronger?
8) If you halved the number of DNA repair enzymes, **would** there **be** more cancers?
9) The digital revolution has greatly increased the use of cameras. If there **is** an optical revolution will there be a similar increase in video use?

C)
1) What would happen if we made a plastic plane?
2) What will happen, if we make a drug for autistic people that increases emotional sensitivity?
3) What will happen if we make silicon chips dosed with germanium?
4) What would happen if we made more efficient DNA repair enzymes?

Unit 22: Numbers (Answers)

A)
1) False (Sixteenth)
2) True
3) False (One hundred and five)
4) True
5) False (Twenty-sixth)
6) False (Two thousand and two)
7) False (Thirty-seven)
8) True
9) True
10) False (One hundred and fourth)
11) True
12) True
13) False (One thousand and one)

B)
1) Twenty-fourth
2) Eight hundred
3) Twelve million
4) Seventh
5) Ten thousandth
6) Twelfth
7) One thousand six hundred
8) Two thousand

C)
1) 93,000,000
2) 10th
3) 50th
4) 3000, 500
5) 150
6) 100
7) 300,000,000
8) 22nd

Unit 23: Numbers II (Answers)

A)
1) True
2) False (*a/one thirteenth*)
3) False (*a/one sixth*)
4) True
5) True
6) True
7) False (*a/one quarter*)
8) True
9) True
10) False (*a/one half*)
11) False (*a/one half*)
12) True
13) True
14) False (an eighth)

B)
1) One and a half.
2) Three and five sixths.
3) Seven and two thirds.
4) Four and seven eighths.
5) Two and three elevenths.
6) One and three sevenths.
7) Two and a fifth.
8) Three and two thirds.
9) Three and a half.
10) Twelve and four fifths.
11) Eleven and six sevenths.
12) Ten and five eighths.

C)

1) 3/4	7) 4 1/5
2) 1/2	8) 3/7
3) 7/8	9) 8 5/9
4) 2/3	10) 20 3/4
5) 12 1/4	11) 1/3
6) 5/6	12) 7 2/7

D)
1) Three times ten to the eight.
2) 2×10^9
3) Ten to the minus fifteen.
4) 10^{-6}
5) Ten to the twenty-four.
6) One point six times ten to the power of twenty-three.
7) 5×10^3
8) 10^{-3}
9) Two times ten to the five.
10) 10^{30}

Unit 24: Elementary Calculations (Answers)

A)
1) Five hundred and sixty-seven plus (*or and/add on/add*) ninety-eight.
2) Eighty-nine minus (*or take away/less/subtract*) seventy-five. (*Or seventy-five subtracted from eighty-nine.*)
3) Thirty-four times (*or multiplied by*) two.
4) Fifty divided by one hundred and eleven.
5) Six times (*or multiplied by*) five is greater (*or smaller*) than four times (*or multiplied by*) three.
6) Nine hundred and sixty-five minus (*or less/subtract*) forty-three. (*Forty-three subtracted from sixty-five.*)
7) Three thousand six hundred times (*or multiplied by*) zero point five.
8) One hundred and ten minus (*subtract/less/take away*) six point two one nine.(*Or six point two one nine subtracted from one hundred and ten.*)
9) One thousand two hundred divided by zero point seven.
10) Two hundred and thirty-four plus (*or and/add/add on*) four is greater than (*or more than*) eighty-nine point seven eight oh.)
11) Zero point two eight five minus (*subtract/take away/less*) forty five plus (*and/add/add on*) nine point three four.
12) Zero point seven eight plus (*and/add/add on*) four point two five plus (*and/add/add on*) six thousand minus (*subtract/take away/less*) four point five.

B)

1) 89 + 5	7) 6000 - 1200
2) 56 – 18	8) 1,000,001 + 1
3) 12 x 30	9) {(12 + 6)/3} - 6
4) 50 ÷ 5 = 10	10) 5.6 + 2.7218
5) 45 ÷ 3	11) 72,000 x 4
6) 200 + 30	12) 1,500,000 - 200

C)
1) + 20
2) - 420
3) + 2
4) ÷ 6
5) x 3
6) 5 ± 1

D)
1) True
2) True
3) False $6 \times 5 < 40$
4) True
5) False G is equal to six point seven times ten to the minus eleven.
6) True
7) True

Unit 25: Ratios, Percentages and Decimals (Answers)

A)
1) False; 2:5 is two **to** five / two **in** five / two **out of** five
2) True
3) False; zero point six four
4) True
5) True
6) False; twelve point zero three
7) True
8) True
9) True
10) False; seventy point seven per cent
11) False; fifteen to two
12) False; thirty-three point five per cent
13) True
14) True

B)
1) Seven point three four two
2) Thirty-two per cent
3) Zero point three one nine ***or*** point three one nine
4) Eleven point one per cent
5) Six to one / A ratio of six to one / Six in one
6) Fifty-four point seven
7) One to five / A ratio of one to five / One in five / One out of five
8) Forty-five point three per cent
9) Zero point nine three five six ***or*** point nine three five six
10) Thirty point zero per cent ***or*** thirty point nought per cent ***or*** thirty per cent exactly
11) Seven to two / A ratio of seven to two / Seven in two
12) Eighty-nine point five per cent

C)
1) 38.5%
2) 0.082
3) 3:5
4) 99.99%
5) 57
6) 7:6
7) 12.03%
8) 3% or 3.0% etc.
9) 1:4
10) 83.279
11) 0.25%
12) 5:3

Unit 26: Units and Dimensions (Answers)

A)
1) nanometres
2) kilometres
3) gigahertz
4) kilograms
5) millimolar
6) picomolar

B)
1) pM
2) ml
3) µg
4) kg
5) pl
6) km

C)
1) gallon
2) cubic feet
3) ounce
4) quarter
5) dram
6) yard
7) mile
8) acre
9) cubic inch
10) inch
11) pint
12) gram
13) ton
14) stone
15) foot

D)
1) False (milligram)
2) True
3) True
4) True (short ton or stone)
5) True
6) True
7) True
8) False (qr)
9) True
10) True
11) True
12) False (centimetre)
13) True
14) False (g)
15) True
16) True
17) True
18) False (oz)

Unit 27: Units and Dimensions II (Answers)

A)
1) kilojoules per second
2) ohms per metre
3) moles per litre
4) litres per metre
5) millimetres per hour squared
6) coulombs per kilogram
7) moles per metre cubed (or moles per cubic metre)
8) $Nm^{-2}kg^{-2}$ newtons per metre squared (or square metre) per kilogram squared
9) kilogram per metre cubed (or kilogram per cubic metre)
10) kilojoules per mole
11) watts per square metre (or watts per metre squared)
12) coulombs per metre

B)
1) kW/m or kWm^{-1}
2) µmol/l or $\mu moll^{-1}$
3) cm/h^2 or cmh^{-2}
4) Ω/cm or Ωcm^{-1}
5) J/mol or $Jmol^{-1}$
6) g/cm^3 or gcm^{-3} or g/ml or gml^{-1}
(Informally: g/cc)
7) C/m^2 or Cm^{-2}
8) J/kg or Jkg^{-1}
9) mol/m^3 or $molm^{-3}$
10) MJ/s or MJs^{-1}
11) lbs/in^2 or $lbin^{-2}$
12) C/m^4 or Cm^{-4}

C)
1) Correct
2) Incorrect (revolutions **per** minute)
3) Correct
4) Correct
5) Incorrect (47 metres **per** second)
6) Incorrect (1.1 grams per m**l**)
7) Correct
8) Correct
9) Correct
10) Incorrect (disintegrations **per** second is a curie)

D)
1) metres per second squared
2) electron volts
3) newtons per square metre
4) joule seconds
5) counts per second

Unit 28: Units and Dimensions III (Answers)

A)
1) ohms per second
2) kilojoules per metre
3) moles per decimetre cubed/cubic decimetre
4) volts per metre
5) teslas per metre squared/square metre
6) microfarads per kilogram
7) lux per metre cubed (or per cubic metre)
8) newtons per metre squared per kilogram squared
9) webers per millimetre cubed (or per cubic millimeter)
10) kilojoules per square metre (or per metre squared)
11) milliwatts per centimetre squared (or per square centimetre)
12) coulombs per metre

B)
1) Fm^{-2}
2) MJs
3) μTm^{-1}
4) mΩnm
5) srs^{-1}
6) Jm^{-1}
7) $mAcm^{-1}$
8) $nmol\mu l^{-1}$
9) lxs^{-1}
10) ccdm
11) $lbsin^{-2}$
12) mph (or mph)

C)
1) Correct
2) Correct
3) Incorrect (kilojoules per gram)
4) Incorrect (kilometres per hour)
5) Correct
6) Incorrect (101 kPa)

7) Incorrect (9 K)
8) Correct
9) Correct
10) Incorrect (50 Hz or 50 cps)
11) Correct
12) Correct

Unit 29: Time (Answers)

A)
1) The plane will be taking off at quarter past seven in the evening.
2) I logged on to the internet at half past two.
3) The automatic safe in the bank will open at nine o'clock in the morning.
4) There will be a meeting for all the medical consultants at four o'clock in the afternoon.
5) A Mondax pharmaceutical representative will arrive at three o'clock.
6) The project manager wants us in his office by one o'clock in the afternoon.
7) The forecast is for rain by nine o'clock in the evening.
8) You have an interview with a scientific journalist at half past four in the afternoon.
9) At quarter to nine in the morning we will take the coach to the conference.

B)
1) 3.20
2) 1.15
3) 12.30
4) 6.57
5) 8.30
6) 2.35
7) 12.08
8) 11.15

C)
1) The pathologist arrived at fifteen hundred hours.
2) We will arrive at the laboratory at nine twenty.
3) The storm started at nine twenty-seven.
4) I am free at one o'clock.
5) The forensic scientist arrived at the court at two.
6) The research and development division closes at half past five.
7) The DVD recorder will switch on at six forty-five.
8) The lecture on radiation safety ends at eleven thirty.

Unit 30: Dates (Answers)

A)
1) The tenth of the eleventh ninety-three
2) The first of April nineteen eighty-four
3) The eleventh of May two thousand and one
4) The third of the fifth ninety-seven
5) The twenty-second of February nineteen fifty-three.
6) The twenty-fourth of the first ninety-eight.
7) The seventeenth of November nineteen eighty-nine.
8) The seventh of March nineteen ninety-five.
9) The thirty-first of the twelfth seventy-two.
10) The twenty-fourth of June nineteen sixty-four.

B)
1) 23rd August 1980
2) 1/3/84
3) 17th September 2010
4) 29th February 2012
5) 2/3/85

C)
1) January
2) in
3) tenth
4) sixty-four
5) Friday
6) on

D)
1) The term (semester) starts on the fourth of October.
2) Newton stated the Law of Gravitation in sixteen eighty-five.
3) The plane will be ready for use in July.
4) We will next see Halley's comet on the first of January two thousand and sixty-two.
5) We expect the delivery of the new computers on the twelfth of September.
6) The expedition to Kenya will begin on the seventeenth of August two thousand and fifteen.
7) The last Californian earthquake was on the first of October nineteen ninety-three.

Unit 31: Time and Dates (Answers)

A)
1) IBM plan to buy Apple in two thousand and twenty-one (2021).
2) It will take a fortnight to grow the cells.
3) I think that this geological structure took a *(or **one**)* millennium to develop.
4) The birds will migrate south in four months.
5) The Voyager Spaceship lands on Uranus next year.
6) It will be a decade before we have a solution to the AIDS.
7) The drug tests will be completed in two weeks.
8) The slides that you need for your presentation will be ready in two days. *Or*
The slides that I need for my presentation will be ready in two days.

B)
1) Did you take any notes **during** (or *in*) the lecture or were you too busy listening?
2) How long do you have to centrifuge the sample **for**?
3) Doctor's often find tumours **during** (or *in*) routine examinations.
4) **During** (or *in*) the thunderstorm the wind speed reached 70 mph.
5) February had 29 days in 1992, 1988, 1984, so its clear that February has 29 days **every** four years.
6) I have bought a RISC based portable computer that will last me **for** at least 5 years.
7) The effects of steroids last **for** many months after you stop taking them.

C)
1) The plant must be watered daily (*or* at daily intervals).
2) I log onto the internet weekly (*or* at weekly intervals).
3) I'm taking cocaine two hourly (*or* at two hourly intervals).
4) I have to apply for a grant yearly (*or* annually/ at yearly intervals).
5) The centrifuge is serviced five yearly (*or* at five year intervals).
6) I supervise practical work monthly (*or* at monthly intervals).

Unit 32: Money (Answers)

A)
1) If the exchange rate rises above five dollars to the pound, we will not be able to afford to go on holiday.
2) Thirty-four euros to five yuan.
3) It is predicted that in two years the exchange rate will be one dirham to one rupee.
4) The advertised exchange rate was a hundred cedis to the pound, however, we received two hundred cedis for every pound.
5) If the euro is devalued then the pound to dollar exchange rate will rise above five pounds to the dollar.
6) Ten dinars to one yen.
7) The ruble has become a more stable currency; its exchange rate of twelve rubles to the pound has been constant for a year.

B)
1) Two hundred and three dollars only———
2) Eighty-five pounds only———
3) One (or a) thousand rubles only———
4) Nine hundred and fifty eight dollars and thirty-four cents only———
5) Seventy-six pounds and fifty pence only———
6) Eight hundred and ninety-five rupees only———

C)
1) $2 = £1
2) $6 : $2
3) R8.75 : £1
4) Rs20 : $1

5) £1 : $37.40

Unit 33: Singular and Plural (Answers)

A)
1) Laminae
2) Larynges
3) Perinea
4) Phenomena
5) Loci
6) Meioses
7) Syllabi
8) Psychoses
9) Bronchi
10) Peritonea
11) Algae
12) Metropolises
13) Protons
14) Paralyses

B)
1) Bacillus
2) Spectrum
3) Abscissa
4) Vortex
5) Focus
6) Emphasis
7) Arthritis
8) Stratum
9) Genus
10) Species
11) Nebula
12) Oasis
13) Neurosis
14) Radius

C)
1) There are twelve thoracic vertebrae.
2) Calculate the velocities of the three projectiles.
3) We need to harvest three ova for in vitro fertilization.
4) Hypnosis is a recognized psychological therapy.
5) How many protons were released as a result of each nuclear reaction?
6) Use the electron microscope to measure the length of each staphylococcus.
7) If we carried out a curriculum review then we would have to change all of the syllabi. *(**Syllabuses** is also an acceptable answer).*
8) Considering wave-particle duality, can you calculate the equivalent mass of a photon.
9) These conjoined twins have two pubic symphyses.
10) The mathematician drew a matrix as part of his proof.

Unit 34: Approximations (Answers)

A)
1) i)
2) ii)
3) i)
4) i)
5) ii)
6) ii)
7) Same accuracy
8) ii)

B)
1) Approaching, nearly, almost or approximately.
2) Exactly.
3) Almost, about, approximately or nearly.
4) Very approximately.
5) More or less.
6) Coming up to or going on for.
7) Nearly, approximately, about or almost.
8) Precisely.
9) Going on for or coming up to.
10) Roughly.

Unit 35: Direction (Answers)

A)
1) The train was moving in a western direction.
2) New York is east of Los Angeles.

3) London is 3175 miles north of Accra.
4) The wind is blowing in a southwestern direction.
5) The electron moved towards the electric field.
6) The hurricane is coming from the north.
7) The Radiology department is in the east wing of this hospital

B)
1) Osmotic pressure occurs because water tends to move from its initial compartment **towards** the higher solute concentration.
2) To reach the conference, you will have to drive south for 50 miles and enter the city centre. The conference venue is 100 metres north **of** the hospital.
3) The earth rotates around the sun because of a gravitational force **attracting** the Earth to the Sun.
4) The explosion propelled fragments **outwards** for 100 metres.
5) Your compass should always point **towards** magnetic north.
6) The Maglev trains work because the trains are **repelled** by the tracks and so float.
7) The ship is sailing in a southwestern **direction**.
8) Blood flows **up** from the heart into the aorta and then **down** to the iliac vessels. *Or*
Blood flows **upwards** from the heart into the aorta and then **downwards** to the iliac vessels.

Unit 36: Movement (Answers)

A)
1) What happens if you turn the dimmer switch clockwise?
2) Why can he only turn his head by +/- 30 degrees?
3) How many times does the Earth rotate in a week?
4) When did he twist his back? *Or*
How did he twist his back?
5) What will happen if you keep driving in a straight line?
6) Does a bee move randomly from one flower to another?
7) What does circling your answer mean?
8) Flying in a straight line, how long does it take to reach London? *Or*
If you fly in a straight line, how long does it take to reach London?

B)
1) He fired the rifle from a distance of 1000 metres and yet he still **hit** the centre of the target.
2) Does light travel in a **straight** line?
3) The bull **collided** with the matador and killed him.
4) Does an electron rotating around a nucleus **describe** a circle?
5) The centrifuge is **spinning** at a rate of 2000 rpm.
The centrifuge is **rotating** at a rate of 2000 rpm.
6) At first sight the ants look as though they are moving **randomly**, but if you watch carefully you can see that they have a purpose.
7) This hard disk **spins** at 7200 rpm, however I think that I would prefer a fast flash drive.
8) The faster the spines of the windmill rotate, the greater the amount of electricity that the windmill produces. There is no wind now so the spines are not **rotating**. (Best answer).

Unit 37: Comparisons (Answers)

A)
1) This television show is **marginally less** popular with social class 1.
2) Copper has a **40% larger** thermal conductivity **than** aluminium. *Or*
Copper's thermal conductivity is **40% larger than** aluminium's.
3) No, his blood pressure is **85% of** the average blood pressure.
4) The modifications to the engine made the car **noticeably faster**.
5) Yes the addition of the enzyme made reaction A **25% faster** than reaction B.
6) The performance of this group of students is **considerably** better than the last group. *Or*
This group of students' performance is **considerably better** than the last group's.
7) Yes, patients with achondroplasia are **far smaller** than average.
8) Yes, the probability of winning the National lottery is **as likely as** being struck by lightning.

B)
1) Species A seems bigger than species B. In fact, if you compare the average heights of the two species you can see that, species A is significantly **taller** than species B. This confirms my first impression.
2) I was hoping for definitive results from this experiment. However the data are not **as** conclusive **as** I expected.
3) The patient's urine output should be about 25 millilitres per hour. In the last hour it was only 5 millilitres.

This is far **less** than expected.

4) If you increase the speed of the processor and increase the size of the LCD screen, you will need a 100% greater power supply. My calculations confirm that the rechargeable battery will need to produce a power output that is **twice** as large as the previous model.
5) A diet low in fresh fruit and vegetables greatly increases the chance of a range of cancers. For example, the probability of developing rectal carcinoma is 50% more in a low fruit and vegetable diet **than** in a high fruit and vegetable diet.
6) The maximum velocity of this train is only 85% **of** the predicted value.
7) To get a safety certificate, the material needs to be sufficiently flame retardant. Can this synthetic material resist flames **as** much as is required?

Unit 38: Processes (Answers)

A)

1) Respiration **starts with** glycolysis and then continues with the Kreb's Cycle. *Or*
   Respiration **begins with** glycolysis and then continues with the Kreb's Cycle.
2) This chemical reaction is initiated **by** a photon of light.
3) The movement of this engine is **triggered by** the combustion of petroleum. (Best answer.)
   The movement of this engine is **initiated by** the combustion of petroleum.
4) I started my presentation promptly **at** 1pm.
5) The morphine pump is **triggered** every time the patient presses the button. (Best answer.)
   The morphine pump is **started** every time the patient presses the button.
   The morphine pump is **initiated** every time the patient presses the button.
6) The emotion of anger begins **in** the brain with a release of noradrenaline. However, it is not clear what **triggers** this release.
   (Best answer. **Starts**, **initiates** or **begins** can be used in place of **triggers**.)
7) A biomechanical analysis of this tennis player's serve, **begins at** the top of the swing.
   A biomechanical analysis of this tennis player's serve, **starts at** the top of the swing.

B)

1) When did the pharmaceutical company start to sell the new painkiller?
2) What triggered rainfall over the east coast of India?
3) Why does coagulation of the egg start? *Or* Why does coagulation start?
4) What acts as a nidus to initiate the crystallization of a urinary calculus?
5) What begins the digestion of protein in the stomach? *Or* Where does pepsin begin the digestion of protein?
6) When was the download of this file triggered by the user?
7) How does the protocol start? *Or* What does the protocol start with? *Or* With what does the protocol start?
8) Where did her journey around the world begin?
9) Did he start his proof by making assumptions that were not justified?
10) With what (muscle) does a smile begin? *Or* How does a smile begin?

Unit 39: Processes II (Answers)

A)

1) Initially, the bullet is accelerated to a muzzle velocity of 975 m/s and then **maintains** a velocity of over 200 m/s for the first 100 m.
2) The crack in the crystal begins at point A, then **continues** to point B and finally ends at point C.
3) I **stopped** reading the article and ordered one of the references. Then I **continued** reading the article. *Or*
   I **stopped** reading the article and ordered one of the references. Then I **finished** reading the article.
4) The thermostat will ensure that the waterbath is **maintained** at 37 °C. However, after an hour the reaction will be stopped by turning off the thermostat and by adding 2 mM NaCl.
5) At 10 pm I will **stop** working on this presentation. However, I don't just want the work to be halted, I want the presentation to be properly **completed**. *Or*
   At 10 pm I will **stop** working on this presentation. However, I don't just want the work to be halted, I want the presentation to be properly **finished**.
6) If the plane **maintains** its current speed, it will arrive at Toronto airport in 3 hours.
7) If the plane **continues** to fly in this direction, it will arrive at Toronto airport in 3 hours.

B)

1) The metabolism of bilirubin begins **in** the liver, **continues** through the length of the small intestine and is completed **in** the large intestine.
2) Before you can start a building project of this magnitude, you must get funding. Then you must make sure that the project begins on time, **continues** within its financial limits and is **completed** promptly.
3) Cancer formation in the lower oesophagus begins **with** metaplasia, then **continues** with dysplasia and ends **with** an invasive tumour.
4) First the reactants are adsorbed onto the surface of the catalyst. Then the reaction continues as one of the reactant

achieves a high energy state whilst the other reactant **maintains** its original state. The proximity of the two reactants allows the reaction to occur. Finally, the reaction is **completed** and the products are released.

5) This presentation must begin at 9 am and **finish** (*or* **end**) at 10 am. The next speaker **starts** (*or* **begins**) at 10.05 am.

Unit 40: Probability (Answers)

Scenario 1) A statistician talking to a client.
Statistician: I can offer a range of services on a private basis.
Client: That sounds good. Though what I really need help with, is preparing my statistical data for publication.
Statistician: No problem. I can calculate the statistical significance of your results; in other words how **likely** you were to have achieved the same results by random chance.
Client: Excellent. I am **certain** that you will give me the help I need. (Or **sure** or **definite**)

Scenario 2) A chemistry lecturer talking to an undergraduate student.
Student: I was interested in your description of the factors that control the reaction rate.
Lecturer: The most important consideration is the proportion of molecules that are above the activation energy for the reaction. Basically, the greater the proportion of such high energy molecules, the more **probable** it is that the reaction will occur. (Or **likely**)
Student: I think I understand. You are saying that high energy molecules are more **likely** to react.
Lecturer: That is essentially correct. It also follows that if two molecules have the same energy, the probability that the reaction will occur is equal.

Scenario 3) A zoologist giving a commentary to a television program.
"The mating ritual of these animals is prolonged and intricate. At first, the female looks upon the male as an unlikely or **improbable** mate. However, after two hours, she sees him as a possible mate. After another hour she sees him as a probable or **likely** mate. If the ritual continues for four hours then the suitor becomes a sure or **certain** (or **definite**) mate."

Scenario 4) A doctor talking to a patient about the risk of lung cancer.
Doctor: I cannot give you an absolute measurement of your chances of getting lung cancer, but I can tell you some of the factors that increase the risk.
Patient: What factors?
Doctor: Smoking for example. It makes the **likelihood** (or **possibility** or **probability**) of getting lung cancer much greater.
Patient: Will I be sure to get lung cancer if I smoke?
Doctor: No. But the chances will go from an improbability to a possibility.
Patient: What is the best way to be **sure** (or **certain** or **definite**)?
Doctor: The only way to be certain is to perform a lung biopsy.

Scenario 5) Two doctoral scientists talking about a colleague during a coffee break.
Scientist 1: He has applied for a prestigious research grant and fellowship.
Scientist 2: Yes, I have heard that he is **certain** (or **sure** or **definite**) that he will succeed. However with a 10/1 applicant to appointment ratio, I feel that he is being optimistic.
Scientist 1: The numbers suggest that it is possible that he will succeed. However if you look at the figures objectively, it is **improbable** (or **unlikely**) that he will be successful.
Scientist 2: What if he has the support of the Professor?
Scientist 1: That would change everything.

Scenario 6) An actuary explaining his job.
First of all, you have to understand that insurance companies are businesses that exist to make a profit. They must set their charges and rates at a level that is **certain** (or **sure**) to make money. Actuaries have the mathematical skills in probability and statistics to ensure that this is the case. For example, if you want to sell a 25 year old middle class male life insurance, you must calculate the probability of him dying in the next 40 years. Hopefully, the chances of this happening are small. It should be an **improbable** (or **unlikely**) event but a good actuary can make a precise calculation of the expectation.

Unit 41: Temperature and Temperature Changes (Answers)

A)

1) The protocol states that the solution should be **heated** to 100°C for 30 minutes.
2) Dry ice (frozen carbon dioxide) can **sublime** from solid to liquid when **warmed**. *Or*
   Dry ice (frozen carbon dioxide) can **sublime** from solid to liquid when **heated.**
3) **Condensing** the organic solvent from gas to liquid is the final stage in the production process.
   Liquefying the organic solvent from gas to liquid is the final stage in the production process.
4) The water on the surface of the building **freezes** in very cold weather. *Or*
   The water on the surface of the building **solidifies** in very cold weather.
5) This hypothermic patient must be **warmed** gradually to a normal body temperature.
6) This purification process relies on **heating** the gold until it melts.

7) Did you know that when bears hibernate in the winter their core body temperatures **cool** to 25°C?

B)
1) What happens if the beverages are lukewarm?
2) How is snow formed?
3) After the thermostat was switched off at 10pm, how much did the room cool by? *Or*
What happened after the thermostat was switched off at 10pm?
4) What caused the extensive burns on this individual? *Or*
How were the extensive burns on this individual caused?
5) In a flame, what colour do calcium salts burn? *Or*
What colour do calcium salts burn in a flame?
6) What happens to the lubricating oil if the engine becomes too hot? *Or*
What if the engine becomes too hot?
7) How does perspiration cool down the body?
8) What happens to the albumin solution as it is heated?
9) Do bacteria proliferate more slowly in cool liquids or warm liquids?
10) When does this metal become a superconductor? *Or*
At what temperature does this metal become a superconductor?
11) What happens to the reaction rate if you increase the temperature of this biological system by 10°C?
12) After the plane crash how were the burnt bodies identified?

Unit 42: Presentations (Answers)

A)
1) Having discussed the costs of the suggested changes, I would now like to move **on to** a consideration of the benefits.
2) So, let's begin **with** a description of the normal anatomy of the human shoulder. Afterwards, we can consider some of the common shoulder pathology.
3) Firstly, the current form of the periodic table will be discussed. Secondly, the solid non-metals will be examined. **Thirdly**, the gaseous non-metals will be examined. **Fourthly** and finally, the rest of the elements will be considered.
4) I'm going to **talk** about the causes, mechanisms and risk factors for diabetes mellitus.
5) The **purpose** (*or* **aim**) of this presentation is to review the latest developments in subatomic physics.

B)
1) Correct
2) Incorrect. Firstly, we will address the regional politics, secondly, the regional economics and **thirdly**, the regional demographics.
3) Correct
4) Incorrect. **Right** then, the purpose of this presentation is to review recent research.
5) Incorrect. So, let's **begin** with a discussion of the increased risk of heart attacks in migraine sufferers.
6) Correct
7) Correct
8) Correct
9) Incorrect. The aim **of** this talk is to review progress in alternative energy sources.
10) Correct
11) Correct

Unit 43: Presentations II (Answers)

A)
1) Now I **would** like to consider the demographic of this disease. **Looking** at Table 1 we can see an increasing incidence.
2) To **summarize** our findings, no statistically significant data have been elicited, however several trends were noted. *Or*
To **recapitulate** our findings, no statistically significant data have been elicited, however several trends were noted.
3) **Finally**, I would like to thank you for listening. Are there any questions?
4) As you can see, this table summarizes the raw morbidity and mortality data. **Now** I would like to analyze the data in more detail. *Or*
As you can see, this table summarizes the raw morbidity and mortality data. **Next** I would like to analyze the data in more detail.
5) **Could** I ask you to look at Figure 7? This is a plot of reaction rate versus time.
6) So how can we synthesize this dye? If you **turn** to the flow chart, then the answer becomes clear.
7) To **sum up**, it is clear that despite the theoretical progress that has been made in the development of nuclear fusion, we are many years from building a working reactor.

B)
1) Incorrect
To sum **up**, the latest quality control results for this electronic component are entirely unsatisfactory.

2) Correct
3) Correct
4) Correct
5) Incorrect
Looking **at** table 3, it can be seen that there is a great deal of variation in the relative financial success of the biological and physical science research divisions.
6) Correct
7) Incorrect. Ladies and gentlemen, would you now turn **to** Figure 4.
8) Correct
9) Correct
10) Incorrect
In conclusion, I would like to thank you for listening. Are there any questions?

Unit 44: Presentations III (Answers)

A)
1) I *believe* that we should extend the study to include rarer consequences.
2) In the 1960s Bradford Hill *postulated* criteria for disease causation in epidemiology. *Or*
In the 1960s Bradford Hill *suggested* criteria for disease causation in epidemiology.
3) Because of these advantages, I *think* that the silicon transistors in your computer will be replaced in ten years by transistors based on carbon nanotubes.
4) *We feel* the evidence for any damage to biological tissue is weak.
5) We *theorize* that it is the overall pattern of the diet rather than any one component that is important in preventing depression.
6) I *propose* that we will be able to double the processing power of the computer by using the Graphics Processing Unit more efficiently.
7) *A I see it*, the automated synthesis of such sugars is now possible.

B)
1) We *theorize* that global warming is caused by carbon dioxide accumulation in the atmosphere but the evidence for this hypothesis is not conclusive.
2) Because of the evidence that I have presented, I strongly *believe* that our theory is the most accurate description of drug-receptor interactions.
3) In the 17th century Sir Isaac Newton *postulated* and published his three laws of motion.
4) Now, *as I see it,* there is very little preventing us from using microbial respiration to generate electricity on a large scale!

Unit 45: Graphs (Answers)

A)
1) We can see from the graph that there was a very large increase in velocity with time. In fact, the velocity shot up/jumped up over the period of this experiment.
2) Considering the world population between 1820 and 1920, the total number of people did not change much. The number was stable/steady/constant at about 1 billion.
3) The temperature change occurred in two phases. Initially the temperature plunged/plummeted by 70% and then decreased/fell/declined/went down/dropped by 20%.
4) The cost of a fluorimeter has not been constant. In recent years there has been a significant variation/fluctuation/oscillation in prices.
5) The patient's measured heart rate was constant/stable/steady for six hours. Then it suddenly jumped up/shot up, so I suspect an infection.
6) As the pressure on the fixed mass of gas increases, the volume plunges/plummets. The findings are consistent with Boyle's law.

B)
1) There was an increase in the number of colony forming units of 62.
The number of colony forming units increased by 62.
2) There was a rise in his body mass index of 10.
His body mass index rose by 10.
3) There has been a decrease in humidity of 20%.
The humidity has decreased by 20%.
4) There was an increase in their energy of 90 MeV.
Their energy was increased by 90 MeV.
5) There was a drop in sound level of 80 decibels.
The sound level dropped by 80 decibels.
6) There will be a climb in the maximum capacity of a hard disk of 98 TB.

The maximum capacity of a hard disk will climb by 98 TB.
7) There was a decrease in speed of 40 miles per hour.
The speed decreased by 40 miles per hour.

Unit 46: Graphs II (Answers)

A)
1) The rate of newspaper production increased dramatically.
2) There will be a gradual decrease in fluorescence of this compound each year.
3) The value of these cars has declined slightly.
4) There has been a constant fall in the income from these shares each year.
5) Did the temperature of the solution increase rapidly?
6) There was a sharp rise in the urine output of the patient two days after the operation.
7) Did the solubility decrease strikingly as the temperature decreased?
8) The pressure declined regularly as the plane rose.
9) After sunrise, there is a rapid increase in the number of photons reaching each leaf.
10) EEG activity decreased dramatically at 8pm; we believe that this was when he died.
11) The conductivity of the amalgam increased greatly after raising the copper content.
12) There was a substantial fall in traffic flow when the parallel road was closed.
13) There has been a steady rise in the incidence of Chlamydia in the under-25s.
14) What would happen if the sale price of the Nikon microscope fell sharply?
15) There was a dramatic rise in the ratings for this television show after the lead actor was changed.
16) There has been a regular drop in the inflation rate in this country every year for the last 10 years.
17) It is predicted that in three years time, flash drive capacity will increase rapidly.
18) There is an initial quick fall in the temperature of a hot liquid. *Or*
There is a quick fall in the temperature of a hot liquid initially.
19) There is a steady rise in the melting point of a hydrocarbon as the carbon chain length increases.
20) There is not a sharp increase in wind resistance with increasing velocity.
21) Her blood pressure dropped substantially immediately after the operation.
22) The Doppler ultrasound showed that blood flow decreased significantly at the carotid arteries.
23) Is there a dramatic increase or a dramatic decrease in the velocity of sound in cold air?

Unit 47: Graphs III (Answers)

1) Obviously there is an **association** between the popularity of a website and the number of downloads.
2) The line graph seems to show that **exposure** to radiation increases the risk of lung cancer.
3) Do you think that mobile phone use **causes** cancer or is associated with cancer?
4) Was a pie **chart** the best way to present your data?
5) The histogram shows that the frequency of herpes infections increases with age. What if age is just a **confounding** factor?
6) Stefan's law states a strong **correlation** between the temperature of a black body and the total energy radiated. The relationship is elegantly displayed on this line **graph** of temperature against radiation.
7) If you compare last year's pie **chart** with this year's **pie chart**, there is a clear drop in the proportion of the company's profits spent on this division.
8) Hooke's law states that the force applied is directly proportional to the extension. Are the **results/data** presented on the graph consistent with this?
9) Would it be better to present the data on an area graph or a stack **bar graph/column graph**?
10) If you plotted a **graph** of microchip processor speed against the year, you would see a line going up rapidly.
11) Can we use the line **graph** of inflation rate against time, to predict next year's inflation rate?
12) Our results are **statistically significant**. In fact there is a 99.9% chance that they did not occur by chance.
13) The data do not support rejection of the **null** hypothesis. However a larger sample is needed.
14) The bar **chart/graph** shows that the median group for male height is 175 cm – 180 cm.
15) If your results satisfy the Bradford-Hill criteria then **causality** is more likely than just an association.
16) He is one of the few remaining people that believe smoking does not **cause** cancer.
17) Your graph shows there is a **correlation** between individual affluence and individual longevity.
18) The slope of your graph of distance against time represents speed. Is there a statistically **significant** change in the speed?
19) What type of **graph** would be best for demonstrating Michaelis-Menten enzyme kinetics?

APPENDIX 1

MEDICAL SPECIALITIES

The table below contains a list of the medical and surgical disciplines, together with a list of specialists (doctors of that discipline) and corresponding adjectives:

Discipline (Speciality)	Specialist	Adjective
Allergology	Allergologist	Allergological
Anaesthetics	Anaesthetist	Anaesthesiological
Andrology	Andrologist	Andrological
Angiology	Angiologist	Angiological
Cardiology	Cardiologist	Cardiological
Dentistry	Dentist	Dental
Dermatology	Dermatologist	Dermatological
Endocrinology	Endocrinologist	Endocrinological Endocrinologic
Epidemiology	Epidemiologist	Epidemiological
Gastroenterology	Gastroenterologist	Gastroenterological
General Practice	General Practitioner	______________
Geriatrics	Geriatrician	Geriatric
Gerontology	Gerontologist	Gerontological
Gynaecology	Gynaecologist	Gynaecological
Haematology	Haematologist	Haematological
Hepatology	Hepatologist	Hepatological
Immunology	Immunologist	Immunological
Laryngology	Laryngologist	Laryngological
Medical Genetics	Medical Geneticist	Medical Genetic
Microbiology	Microbiologist	Microbiological
Nephrology	Nephrologist	Nephrological
Neurology	Neurologist	Neurological
Neuroradiology	Neuroradiology	Neuroradiological
Neurosurgery	Neurosurgeon	Neurosurgical
Obstetrics	Obstetrician	Obstetric
Oncology	Oncologist	Oncological
Ophthalmology	Ophthalmologist	Ophthalmological
Optometry	Optometrist	Optometric Optometrical
Orthodontic	Orthodontist	Orthodontic
Orthopaedics	Orthopaedist	Orthopaedic Orthopaedical
Osteology	Osteologist	Osteological
Otolaryngology	Otolaryngologist	Otolaryngological
Otology	Otologist	Otological
Paediatrics	Paediatrician	Paediatric
Pathology	Pathologist	Pathological

Physiotherapy	Physiotherapist	Physiotherapeutic
Podiatry	Podiatrist	Podiatric
Proctology	Proctologist	Proctological
Psychiatry	Psychiatrist	Psychiatric
Pulmonology	Pulmonologist	Pulmonological
Radiology	Radiologist	Radiological
Rheumatology	Rheumatologist	Rheumatological
Rhinology	Rhinologist	Rhinological
Serology	Serologist	Serological
Surgery	Surgeon	Surgical
Toxicology	Toxicologist	Toxicological
Traumatology	Traumatologist	Traumatological
Urogynaecology	Urogynaecologist	Urogynaecological
Urology	Urologist	Urological

Be careful to note the following:

A) In American English –ae is replaced with -e, for example:

H**ae**matologist – British English
H**e**matologist – American English

P**ae**diatrics – British English
P**e**diatrics – American English

B) Compound terms not in the above table, can be derived using the examples above:

Pathology, Pathologist, Pathological, can be used to form **Paleo**pathology, **Paleo**pathologist and **Paleo**pathological.

Plastic surgery, **plastic** surgeon and **plastic** surgical can be similarly derived.

APPENDIX 2

SCIENTIFIC AND TECHNICAL SPECIALITIES

The table below contains a list of the scientific and technical disciplines, together with the conventional names for people who are expert in the respective discipline and the corresponding adjectives:

Subject	**Person**	**Adjective**
Anatomy	Anatomist	Anatomical
Anthropology	Anthropologist	Anthropological
Archaeology	Archaeologist	Archaeological
Astrobiology	Astrobiologist	Astrobiological
Astronomy	Astronomist	Astronomical
Astrophysics	Astrophysicist	Astrophysical
Bacteriology	Bacteriologist	Bacteriological
Biochemistry	Biochemist	Biochemical
Biology	Biologist	Biological
Biophysics	Biophysicist	Biophysical
Botany	Botanist	Botanical
Cell Biology	Cell Biologist	Cell Biological
Chemistry	Chemist*	Chemical
Chronobiology	Chronobiologist	Chronobiological
Ecology	Ecologist	Ecological
Electricity	Electrician	Electrical
Embryology	Embryologist	Embryological
Endocrinology	Endocrinologist	Endocrinological
Engineering	Engineer	Engineering
Genetics	Geneticist	Genetic
Geochemistry	Geochemist	Geochemical
Geography	Geographer	Geographical
Geology	Geologist	Geological
Geomorphology	Geomorphologist	Geomorphological
Geophysics	Geophysicist	Geophysical
Hydrology	Hydrologist	Hydrological
Immunology	Immunologist	Immunological
Mathematics	Mathematician	Mathematical
Metallurgy	Metallurgist	Metallurgical
Microbiology	Microbiologist	Microbiological
Mineralogy	Mineralogist	Mineralogical
Molecular Biology	Molecular Biologist	Molecular Biological
Neuroscience	Neuroscientist	Neuroscientific
Nutrition	Nutritionist	Nutritional
Palaeontology	Palaeontologist	Palaeontological
Parasitology	Parasitologist	Parasitological

Pathology	Pathologist	Pathological
Pharmacology	Pharmacologist	Pharmacological
Photography	Photographer	Photographic Photographical
Physics	Physicist	Physical
Physiology	Physiologist	Physiological
Psychology	Psychologist	Psychological
Science	Scientist	Scientific
Statistics	Statisticians	Statistical
Teratology	Teratologist	Teratological
Toxicology	Toxicologist	Toxicological
Virology	Virologist	Virological
Xenobiology	Xenobiologist	Xenobiological
Zoology	Zoologist	Zoological

N.B. In English, "Chemist" also means someone who sells medicines/pharmaceuticals.

www.ingramcontent.com/pod-product-compliance
Ingram Content Group UK Ltd.
Pitfield, Milton Keynes, MK11 3LW, UK
UKHW060016051026
15591UKWH00002B/224
* 9 7 8 0 9 5 6 6 4 4 3 0 5 *